REALITY'S FUGUE

F. Samuel Brainard

REALITY'S FUGUE

Reconciling Worldviews in Philosophy, Religion, and Science

The Pennsylvania State University Press
University Park, Pennsylvania

Names: Brainard, F. Samuel, 1943– , author.
Title: Reality's fugue : reconciling worldviews in
    philosophy, religion, and science / F. Samuel
    Brainard.
Description: University Park, Pennsylvania :
    The Pennsylvania State University Press, [2017]
    | Includes bibliographical references and index.
Summary: "Explores complex questions about the
    nature of reality, philosophy, and religion, and how
    we reconcile our often-conflicting beliefs about
    these questions"—Provided by publisher.
Identifiers: LCCN 2017033218 | ISBN 9780271079318
    (pbk. : alk. paper)
Subjects: LCSH: Reality. | Religion—Philosophy.
Classification: LCC BD331 .B725 2017 | DDC 110—dc23
LC record available at https://lccn.loc.gov/2017033218

Printed in the United States of America
Published by The Pennsylvania State University Press,
University Park, PA 16802-1003

The Pennsylvania State University Press is a member of
the Association of American University Presses.

It is the policy of The Pennsylvania State University Press to use acid-free paper. Publications on
uncoated stock satisfy the minimum requirements
of American National Standard for Information
Sciences—Permanence of Paper for Printed Library
Material, ANSI Z39.48-1992.

Contents

POSTSCRIPTS

fugue (feūg) n [from Italian *fuga*, flight; from Latin *fugere*, to flee] (1597) **1.** A musical composition technique in which one or more themes and their variations are introduced in succession and developed in counterpoint. **2.** A type of psychological disorder.

Preface

This book proposes what I believe is a fresh perspective on the nature of reality, one that may open avenues for reconciling certain competing worldviews in philosophy, religion, and science. Some readers, particularly scholars in philosophy or religion, may find helpful the following comments on this proposal's background and methodology.

Behind my analysis of different accounts of reality is a study of the distinction we make in our everyday lives between what have come to be called the "first-person view" and the "third-person view." I discovered the distinction between these views in a seminar I attended on the philosophy of mind. In many of the articles I read for the course, the two views were used to capture the hard problem of consciousness: the two seemingly impossible-to-reconcile ways one could talk about and explore conscious awareness. On the one hand is consciousness as we know it in our personal experience—how it appears to us from the "inside," so to speak, when we examine the nature of our own conscious awareness. On the other hand is how consciousness appears from the "outside"—how it appears to someone apart from the subject, such as when it's examined in an fMRI or PET scan of a person's brain. From the first-person perspective, we know consciousness firsthand and self-evidently. From the third-person perspective, this is not the case. The presence of consciousness in others is something we infer and can be wrong about; it doesn't have the subjective immediacy of our own conscious awareness.

Prior to attending the seminar, I'd published a book that contrasted three accounts of reality in terms of how each dealt with a certain philosophical puzzle. The distinction between first-person and third-person views offered a way to clarify the puzzle and sharpen this contrast. The two views were closely related to the accounts of reality that I wanted to distinguish and were also relatively obvious and easy to explain, even to someone with little background in philosophy.

My approach to reconciling accounts of reality couples these two views with a reexamination of certain basic philosophical problems, and such an approach may raise questions for some readers. My discussion of contradicting worldviews begins, in effect, from scratch rather than with our contemporary scholarly discourse on the subject, and such a way of beginning may give the impression that the book may be introductory or simplistic. Despite first impressions, my analysis is directed at a wide range of current philosophical issues both within and outside metaphysics. I begin in this way because my approach to these issues requires revisiting certain basic concepts and puzzles that have troubled philosophy for a long time and are staples of its introductory courses. This way of beginning also makes my arguments regarding the nature of reality as clear as possible to any interested reader.

My approach may also seem presumptive—as if I've assumed that accounts of reality in science, philosophy, and religion can be reconciled without considering those who argue otherwise. What of philosophers like Foucault, Derrida, Gadamer, Habermas, Rorty, or the many others who have shown how metaphysical accounts are deeply and seemingly inextricably embedded in culture, language, and history? What of those like Richard Dawkins, Stephen J. Gould, or Auguste Comte who claim that religion and science make incompatible claims, have different concerns, or are different stages in our intellectual maturity?

While this book is very much concerned with (and in sympathy with) metaphysical contradictions stemming from differences in cultural circumstances, its primary focus is basic philosophical issues that appear in more than one culture, have never been resolved, and appear as well among the underlying assumptions of empirical science. World religions can be thought of as philosophies in practice, and certain differences in their views of reality provide good examples of such issues. These metaphysical accounts have endured over millennia and continue to resonate with people around the world. When examined comparatively, certain of these accounts illustrate very clearly the philosophical problems I'm interested in—problems we also find in contemporary Western philosophy, including the philosophy of science.

But what, then, do I propose for these more enduring contradictions among accounts of reality, ones that seem to trouble philosophers regardless of cultural context? In fact, I don't think that such contradictions can be resolved in a way that is both comprehensive and logically coherent. This analysis doesn't aim for a metaphysical synthesis that removes the differences in these accounts. Rather, its goal is a better understanding of a particular puzzle that prevents certain conflicting worldviews from being resolved, and of what the source of that puzzle might be.

This goal affects what is and isn't emphasized in my analysis. Because the particular metaphysical puzzle I'm seeking to flesh out lies at the intersection of different accounts of reality, I examine each account principally to highlight the contrasts that interest me. While I make certain points that should interest specialists, they should not expect to find here thorough, nuanced analyses of these philosophical accounts. That is not my intent. My intent is to use the differences in these religious philosophies to shed light on a philosophical puzzle that seems to make it impossible for any one strategy to be entirely satisfactory in all respects, a puzzle that also helps illumine the accounts of reality themselves.

Given this aim, my sources are also somewhat different from what are generally expected in most scholarly works. Much of the information needed to flesh out the accounts of reality highlighted here is readily available in philosophical and other discipline-specific reference materials. While I also use other sources, these secondary sources are particularly appropriate for my analysis. They are as close as our community of discourse comes to texts intended to suggest collective understandings on various subjects.

These methodological considerations notwithstanding, perhaps the biggest difficulty presented by this proposal will be its novelty to most readers. If my analysis of conflicting views of reality has merit, then we are quite likely at the doorway of what may be for many a new universe, one that, like a 3-D movie, doesn't come into focus except in the convergence of differing points of view. If we are too wedded to a particular metaphysical view, then we won't see this universe. Even if our view is that there exists only a diversity of views, we won't see this universe. As suggested by this book's title (and by Douglas Hofstadter in his *Gödel, Escher, Bach*), the universe described here is perhaps best understood as a kind of fugue, a musical composition expressed through many voices. Some of these voices harmonize; some play against each other; over time, allegiances change. And that which excites us, enchants us, challenges us, and gives meaning to these many voices is what is achieved in their interplay, the music that comes into being in the sound they make together.

But of course this is only a metaphor, a trope attempting to clarify what can easily be thought of as just another account of reality, another point of view out of many.

So here is a question, one that runs through this book: How can I, or you, or anyone be just one point of view yet legitimately claim to speak for more than just one point of view? Or, to put it in another perhaps more familiar way, How is philosophy or science possible?

SAMUEL BRAINARD
FEBRUARY 11, 2017

Introduction

I propose in this book a way to reconcile competing answers to certain long-standing fundamental questions about our human nature and the nature of the universe we live in.

What, for example, do we mean when we say that something is "real" or "true" or that something is "alive" or "exists" or has "conscious awareness"? How does our experience of having thoughts and feelings and a mind fit with the nature of matter and with quantum mechanics? What about life's purpose and God? How do we reconcile religion with science? What justification is there to believe in a spirit realm or that we are something besides just our physical bodies?

In the enormously diverse worlds of philosophy and religion, there's little agreement on answers to questions such as these. This book develops a way to understand certain of these different answers as pieces of a larger puzzle, as expressions of a particular mystery of our universe, a paradoxical process basic to the behavior of both living beings and matter.

There are three parts to the development of this proposal. Part 1 introduces the concept of reality as well as philosophical difficulties associated with it that play central roles in our understanding of ourselves and the universe we live in. Also introduced is the distinction between first-person and third-person views mentioned in the preface. Part 2 examines three contrasting accounts of reality and the strengths and difficulties of each approach. Although these accounts are drawn principally from world religions, the philosophical issues examined are ones that humankind has been struggling with for a very long time both inside and outside of religious contexts. Part 3 proposes a way to understand the three accounts as different themes in a single composition. Approached in this way, they would seem to reveal much more about ourselves and the nature of our universe than can be gained through any one account alone.

Much rests on the philosophical issues addressed here, especially as they relate to religion and our spiritual lives. These issues concern not only the nature of reality but also the credibility of various religions and what they tell us about life's meaning and purpose. Religions are, among other things, our philosophies in practice, and how we decide these philosophical issues deeply affects our religious beliefs, spiritual practices, and possibilities for fulfillment in life.

I've written this book as much for myself as for others; I've sought to write the book that I would have liked to have read when I was an undergraduate and struggling in my own life with these basic philosophical and religious issues. As such, this book is not only about reconciling worldviews; it also offers an alternative introduction to philosophy and especially the philosophies underlying world religions.

PART ONE

What Is Real?

CHAPTER ONE

The Predicament

Of Schizophrenia, Religions, and Conflicting Views of Reality

I had a younger brother Bruce who had schizophrenia. He heard voices no one else heard and for many years called these voices real. I have little doubt they were as real to him as my voice was when we were talking together.

Bruce's hallucinations didn't start until he was about thirty. Earlier, when we were in college, as Bruce pursued his interest in mathematics and I an interest in physics, both of us thought of reality as the logically structured, seemingly true-for-all universe that scientists found on the other side of microscopes, telescopes, and other tools of empirical research. With his illness, reality for us continued to mean something like true for all, but it now also had a subjective dimension.[1] Our conversations on these subjects now included the human mind—those aspects of reality on the near side of this research equipment.

When I started writing this book, a frequent topic of our talks was what Bruce at the time was calling his S.E.I.E.W.—his "Self's Enemy Imaginary External World." This imaginary world was his mental copy of reality but was populated with the voices he heard—voices that ordered him about, antagonized him, and were what he thought of as enemies. That he should speak of these voices as imaginary was a major milestone in his life, one that probably would never have happened without the medications he was taking.

My brother's illness brings me to the question of how a person decides between what is and isn't real, what is and isn't true for all. On what basis do we make such judgments? In Bruce's case, even as he got a little better, what he called "real" still often included fanciful circumstances, such as people spying on him or communist takeover plots. He judged whether these circumstances were true for others as well as himself and not inventions of his mind largely on the authority of his inner voices and whatever psychological dispositions they might have expressed. For most of his illness, what he claimed to be real rarely differed very much from what his voices said was so.

I also mention my brother's illness because this book is about religions—philosophy as we find it in practice—as well as about philosophy itself, and many people (such as Freud) claim that religions resemble psychological disorders.[2] To them, the visions and voices that come with certain spiritual experiences have more in common with my brother's voices than with any genuine insight into a higher truth. Faced with a mortal existence perhaps too meaningless or otherwise unsatisfying to cope with in any sane way, religions, in their view, conjure a happier, though illusory, reality.

When I was an undergraduate, I heard a story about an experiment with a Skinner box—one of those cages used in learning research where food would appear if an animal pressed the proper lever. As so often done before, rats were trained to expect to get food when they behaved a certain way. In this case, however, once the rats were trained, the experimenters started feeding them randomly, irrespective of lever pushing. Regardless of what the rats did, sometimes they would be rewarded and sometimes they wouldn't. And when the rats faced a world of random rewards, they developed rituals. The experimenters, so I heard, called these rituals "religious behavior."[3]

Personally, I side with religions against this Skinner box version of religious behavior. I don't think that the happiness religions claim to offer depends on ignorance or delusion about the nature of reality. But I also don't think that religions make their cases very well. Too often, they argue for their versions of truth as if their beliefs were the only right ones and their scriptures the only authorities on such matters. When the sole sources of authority people accept are those within their own religions, these sources begin to look to me very much like my brother's voices. What they say about reality seems to have little basis in what's genuinely true for everyone, including those who believe differently.

In fairness to religions, their different versions of reality seem, on the face of it, impossible to reconcile. Western monotheists—Jews, Christians, Muslims—typically believe that both God and the everyday world of one's sense experience (God's "Creation") are real. God is not a figment of the human mind, nor, typically, are the cars, people, trees, and other things we perceive around us every day. But there are many people who believe otherwise. Buddhists don't believe in God. Nor do they, along with many Hindus, believe that these cars, people, and trees of our everyday experience are entirely real either. For them, the world that we're aware of around ourselves is, in certain basic respects, a creation of our minds; it's not something that exists independently of the way we're aware of it. Then there are the older, indigenous religions throughout the world, such as Shinto of Japan, the popular varieties of Daoism of China, and native American and African shamanic beliefs that have yet other versions of reality. Typically, in religions like these, the everyday world of our sense experience is not thought of as created by our minds as in Hinduism and Buddhism; however, the world is suffused with gods or spirits. Not only are cars and people parts of reality, but so are ancestor spirits or demons or spirit realms or supernatural potencies.[4] And as if this mélange of contending realities were not enough, modern science has introduced its own strange and uncomfortable version—a cosmos of warping and even foaming space-time, strangely behaving subatomic particles, and a universe of such mind-numbing vastness as to undercut, so it seems, any significant purpose or meaning for our local lives here on earth.

So how are we to reconcile these and other versions of reality? What authority, if any, can we turn to for our choice that might be truly neutral and not just the voices of our psychological dispositions?

Science's Limitations

No doubt science offers humankind's best effort to date for discovering pancultural truths about ourselves and the universe we live in. On the basis of its findings have come computers, cell phones, vaccines, pilotless aircraft, a space stations, electron microscopes, and bioengineering. The global success of such technology convincingly demonstrates, at least to me, the legitimacy of science's account of reality.

But science's account, as helpful as it may be, takes us only so far. The concern of science is universal truths, and to this end, it uses logic to sift through the data of experience to find what is universal. But the basis for logic and the nature of universals are among the unresolved issues that have vexed Western philosophy since almost its beginning. To begin with, we don't know why our universe appears to behave logically and why we can use mathematics to describe it. What, for that matter, is mathematics, and exactly what kinds of things does it and doesn't it describe? (Think, in this regard, about the sentiments of a poem, the thrill of hearing certain music, a parent's love for a child.) Furthermore, we also don't know how there can be universals—things that are the same from instance to instance. Why should a photon in some distant galaxy seem to follow the same rules as a photon here on earth? Where does this sameness between the two come from? The objects themselves? The human mind? The nature of the universe as a whole? Then too, does anything truly remain the same over time? Might it not instead be change that's the ultimate nature of all things?[5]

LOGICAL POSITIVISM AND THE LIMITS OF SCIENCE

Can science answer all philosophical questions?

Beginning around 1924, some of the most prominent intellectuals of the time (e.g., Rudolf Carnap, Otto Neurath, and Kurt Gödel) collaborated in an effort to prove that it could. They sought to show that the methods of empirical science could be used to resolve all meaningful philosophical questions and that only questions answerable by these methods were meaningful. These "logical positivists" or "logical empiricists" made significant contributions to philosophy, logic, and mathematics, among other fields, and the movement they started thrived in Europe and the United States for more than a quarter century. But they failed in their efforts to supplant philosophy with science and, in so doing, left a legacy that remains an iconic example of the limits of science.

The movement's defeat came on several fronts. To begin with, they were unable to empirically prove that only scientifically verifiable questions were meaningful; such a proof was outside the scope of empirical science. This failure left the door open for other philosophical concerns outside the scope of science to be meaningful—including those in metaphysics and religion.

They also could not agree on a logically rigorous explanation of how science could prove even our most basic conclusions about the world (that there is a table in front of me, for example). Empirical data have their origin in what scientists observe; that is, empirical data involve sense experience, which is, by nature, specific to the moment. Like anyone else, what a scientist directly experiences at one time and place is never exactly the same as what he or she experiences at other times and places. Theories of science, however, are not specific to the moment; they are part of our public discourse and apply to a variety of circumstances. To show how experimental data (e.g., what one sees with a telescope) verified theories of science (e.g., the rules governing celestial objects), the positivists needed to formulate logical state-ments, or "protocols," that bridged the gap between the sense data of direct experience and the public language of empirical science. After decades of attempts, no satisfactory way to bridge this gap was ever found. By 1960, for this, the aforementioned, and other reasons, logical positivism as a movement had largely faded away.[6]

Some of our most basic beliefs about what is real and what isn't depend on how we answer questions such as these. If one thinks, for example, that the logical, cause-and-effect behavior of nature is the sole agency behind all things, then human life is entirely scripted by the laws of nature and free will is a myth. On the other hand, if one turns to his or her own personal experience of making conscious, voluntary choices and takes that as a basic fact of our universe, then human beings do indeed seem to be agents of nature responsible for at least certain local circumstances of our planet. If, again, one thinks that change and impermanence are the ultimate reality of all things, then you, I, religions, philosophies, and whatever gods or God there may be are likewise just temporary ripples in the great flux of time. But if there are things that endure as permanent threads in the fabric of the universe, then there might be room in our cosmos for enduring selves and souls, eternal truths, and perhaps even spiritual beings.

The point is that the edifice of science rests on an enigmatic foundation that has splintered humankind into different metaphysical camps, a foundation that lies largely outside the domain of investigation for science itself. To resolve matters such as these, we turn to philosophy, which brings us back to where we began, to a diversity of views with no apparent basis for choosing among them.[7]

Three Usual Ways to Resolve Conflicting Views of Reality

So again, how do we make sense out of these different views of reality? We can, of course, presume that there is no sense to be made of them. Diversity in this case becomes evidence of a lack of ultimate answers, or at least meaningful ones. But how, then, do we avoid nihilism, concluding that existence is senseless and useless? What more palatable options might we have?

Apart from abandoning the quest to understand reality, writers on world religions have discussed three usual ways of dealing with philosophical and religious diversity.[8]

First, and most obviously, we can pick one alternative as the right one. This is the "exclusivist" approach, in which we believe that our religion or philosophy is true and contrary views are wrong. Most of those who are religious today fall into this category to one degree or another, with the most exclusivist being those who are the most closed minded about and intolerant of opposing beliefs. The problem with this position is not only the enormous social ill will and conflict it breeds; it also fails to acknowledge that other views are very often equally plausible once one comes to understand them. Many of the differences among religions boil down to little more than cultural or personal preference—which texts should serve as scriptural authority, for example, or how the religious organization should be run. As for truly substantive differences such as those that concern the nature of reality, many reflect philosophical questions that have been around literally for millennia and remain unsettled to this day. In these cases, we are caught on the horns of genuine dilemmas—philosophical mysteries that have no clean "right" or "wrong" answers and leave us with a variety of equally reasonable views. The fact that we may have answers that suit ourselves doesn't necessarily make other choices wrong.

Second, we can claim that all alternatives can be understood in terms of one particular account. This is the "inclusivist" approach, in which we treat all alternatives as legitimate if they are in some sense reasonable but think of one account as more basic and comprehensive than the rest. Differences in views in this case are traced to differences in language or culture or politics or perhaps just human capriciousness. This approach can indeed explain some of the differences, such as those of cultural or personal preference mentioned above. Other differences, however, cannot be so easily explained without doing violence to the individual views. How, for example, do we reconcile the God of monotheism with Buddhism or any other well-thought-out atheism without greatly distorting one or the other's beliefs?

Finally, we can treat the contradictions among plausible accounts of reality as a basic philosophical fact. We can be "pluralists" and accept truths and realities as relative to language, culture, and the like. While this approach offers the most unprejudiced view of different religions and philosophies, it is not a very satisfying approach for practical everyday life. What we want out of our religions and philosophies if they are to be genuinely useful to us are solid, pancultural truths that we can count on and on which we can base our life decisions. Does God exist or not? It can't be both, it seems, so which is it? Is there a human soul or not? An afterlife? To what extent are our lives scripted by natural laws? How we live our lives depends on answers to questions such as these, where we seem forced to choose one view as right and another as wrong.

These, then, are the three usual ways of resolving the differences among religions and philosophies.[9] All of them are ultimately unsatisfactory in one way or another, which raises the question of whether there might be another, fourth option.

This question brings us to fugues and the subject matter of this book.

Fugues and a Fourth Option

A fugue is a musical composition technique marked by a distinctive counterpoint structure. A main melodic theme, its variations, and often counterthemes and their variations are first stated by successively entering parts of the music—"voices" as these parts are called. After the thematic material is introduced (though counterthemes and variations can also be introduced later), the fugue as a coherent whole takes form from an interplay of the voices that has often been likened to a conversation or debate or even an argument.[10] Appreciation of a fugue requires a capacity to hear each voice as it is by itself as well as a capacity to understand how the voices fit together. The fugue's "reality," so to speak, comes in the form of voices that mix and interweave yet remain distinct.

If the music of a fugue can express itself in contrasting voices, even contrasting melodic themes, then surely reality could express itself similarly.[11]

In the following chapters, I describe three very different accounts of reality found in world religions. My story of these accounts skips over, for the most part, those differences easily traced to cultural or historical circumstances—preferences in scriptural texts, forms of worship, or administrative structure, for example. Instead, I focus on certain seemingly intractable differences that lie at the root of our conceptions of reality, whether religious or not.

Although I present these philosophical differences as starkly and as true to their religious contexts as I can, it seems to me that together they give a much more enlightening perspective on reality than any individual view in isolation. Understood as different voices in a fugue, as different themes in a single composition, these three views reveal much more about ourselves and the universe than can be gained with a study of any of them alone.[12]

Another Kind of Fugue

I don't wish to stretch this fugue metaphor too far, yet it's perhaps also instructive that the word has another, very different meaning. In psychology, a fugue is a "dissociative disorder," which is a class of illnesses in which elements of the psyche related to one's identity, sense of self, and memory stop functioning as an integrated whole. People become confused about who they are and where they live and work. Disoriented in identity, they wander away from home to travel for hours or even months, often for great distances.

The psychological fugue suggests another story, one of what happens when we fail to integrate critically important themes of our thoughts and experiences. In this story, our conceptions of reality, truth, and who we are break into discordant pieces. We feel disoriented and no longer at home in the world. Less fugue-like but more disturbing symptoms also come to mind—efforts, for example, to coerce all views of reality into one ideological mold or another and achieve by bullying or perhaps even genocide what was not achieved by reason.

This second story speaks to the urgency of reconciling our different views of reality and the unwelcome consequences of ignoring or failing in this task.

The second story also speaks to the consequences of pursuing this task in disregard of our philosophical history. Earlier I mentioned some limitations of science. Neglecting these limitations has psychological as well as philosophical implications. We pay a psychological price when we presume that science gives us a complete picture of reality and then try to understand ourselves and the world on science's terms alone. If we turn the lens of science on ourselves and our circumstances and are truly honest about what we see (which we rarely are[13]), we find human life to be extremely bleak. Against the backdrop of our physical universe, our lives appear no more significant than vanishingly small and fleeting specks of dust. Focusing in further, we find ourselves to be biological organisms struggling to survive in the face of inevitable death and decay. If we focus in even further, we disintegrate entirely into cells, DNA, atoms, and the biophysical rules that compose us at this level. Nowhere in any of these pictures is life meaningful at all.[14] As one physicist, Steven Weinberg, famously put it, in the field of science, "the more the universe seems comprehensible, the more it also seems pointless."[15]

Those of you who are religious (among others) no doubt disagree with this gloomy picture. It leaves out Allah or God or the Brahman of Hinduism or the Dao of Daoism or the Dharma of Buddhism; it leaves out, you will likely say, the crucial piece that provides life's meaning. But such a response doesn't answer the underlying question of whether religions have something of philosophical value to tell us about ourselves and our universe or whether they are merely psychological coping mechanisms.[16]

This book seeks to show—among other things—the credibility of certain religious accounts of reality and make a case for their philosophical relevance.

Summary

Faced as we are today with diverse views of reality (in the rough sense, for now, of what is true for all regardless of personal perspective), how do we separate what is genuinely real from fiction? Why, especially, should we believe the accounts of reality we find in religions when they often seem better explained as psychological coping mechanisms?

Many believe we should turn to science to arbitrate such issues. But as helpful as science has been, it has limitations; in important respects, questions like the two above lie outside its purview. They fall to philosophy rather than science, which leaves us, again, with the problem of reconciling competing accounts of reality.

Writers have discussed three usual ways of resolving such difficulties. We can decide that one account is right and the others wrong (exclusivism), we can decide that one account encompasses all others (inclusivism), or we can decide that there is no one right or better account of reality (pluralism). All these options are unsatisfactory in one way or another.

This book proposes a fourth option. Much like fugues in music, we can regard contradicting views of reality as different themes in a larger composition, a composition that we understand in terms of the different views that make it up. Like the pluralist strategy, differences in philosophical views are preserved; no single account is favored. Unlike the pluralist strategy, there is a larger composition in which all views play a part and in terms of which the contributions of individual accounts can be judged. I will illustrate this proposal by showing how three contradicting philosophical views found in major world religions might in fact complement each other in fundamental and revealing ways.

Fugue also refers to a psychological disorder in which people lose track of who they are and where they live. To me, this second meaning of fugue suggests the urgency of reconciling our different views of reality and the psychological price we pay for ignoring or failing in this task.

CHAPTER TWO

Two Views of Reality

The Need for New Philosophical Tools

I begin my discussion by accepting certain plausible but contradicting pictures of reality as equally valid.[1] The objective is to show how these pictures might contribute together to our understanding of reality, how they might in fact complement each other much as do voices in a fugue.

It seems to me that such a task will not be accomplished without new philosophical tools, new concepts. I realize that it's a bother to learn new terms, but I see no way around it. It is, after all, our existing concepts that have framed our debates over accounts of reality and left us with no satisfactory resolution.[2]

Before discussing changes to our current vocabulary, however, I need to do some preparation. In particular, I want to distinguish two fundamentally different accounts of reality and explain some of the concepts we currently use to talk about them. I also want to give extended illustrations of each, for which I will turn to Hinduism and Buddhism. Once this groundwork is completed, I'll begin reexamining our vocabulary—especially such concepts as "reality," "mind," "matter," and "consciousness."

Following these first two accounts of reality, I'll discuss a third, equally different account, this one illustrated with certain theist religions (e.g., the monotheisms of Judaism, Christianity, and Islam). This third account is an amalgam of the first two. Particularly significant is how *all* these accounts (including the third) have their strengths and weaknesses and how certain philosophical differences that divide them trace back to one underlying philosophical puzzle.

Once this root puzzle is better exposed, it seems to me that reality comes to look very different than it did before. Contrary philosophical views remain, but they now augment rather than undermine each other. For example, in religion, certain versions of atheism and monotheism now seem to describe reality equally well. In philosophy, Descartes's mind-matter distinction comes to have a very different meaning and significance. In science, the physical characteristics of matter now seem to help us better understand human consciousness.

First-Person versus Third-Person Views

To distinguish between the first two accounts of reality I want to discuss, let's begin with how I've already classified the ways we usually deal with religious and philosophical diversity.

Recall that I divided these strategies into three general categories. Exclusivists presume that some views of reality are right and others wrong, and our task is to decide which is which. Inclusivists see our diversity of views coming principally from differences in language, culture, and the like; for them, the task is to explain how all views might be variations of one view. Both of these first two options presume that there are universal truths to be discovered behind the kaleidoscope of philosophical and religious diversity.

Recall also how the third option, pluralism, differs from the first two. Pluralists presume that there are no universal truths that we can know either because they are beyond human comprehension or because they don't exist. What exists instead are relative truths—truths that arise out of and are anchored in the diverse experiences of various individuals and groups.[3]

The distinction among accounts of reality I'm interested in lies in the difference between the first two options and the last. The exclusivist and inclusivist options are both anchored in universals and the way things are for everyone regardless of personal perspective—the way things are from what is often called the *third-person view*. The pluralist option is anchored in the diversity of human experience. It stresses the personal perspective and the philosophical importance of the *first-person view*, the view of individuals themselves.[4]

However, it is not these two approaches to understanding reality that I want to talk about right now but rather the distinction between the first-person and third-person views associated with them. It's a very basic distinction in our everyday lives but also one with tentacles reaching into a host of troublesome philosophical issues.

Let's look at these two views more closely.

Imagine that you are sitting in a chair reading the book of Genesis in the Bible. No doubt you understand what you are reading in your own personal way. You bring to the reading your own history, your own sense of what various words and phrases mean. You may take the creation story literally or metaphorically, as just literature or myth; what passages convey to you is different from what they may convey to another.

In this sense, we each experience this book in our own personal "first-person" way. What it is for you is whatever you personally are aware of—whatever you, yourself, perceive it and understand it to be.

On the other hand, there is also the book of Genesis that others have read—the one people have talked about, disagreed over, and picked up and read again for centuries. In this respect, the book of Genesis is something much more than or other than your own personal experience of it. It is something that exists for others to experience as well. The book is a shared feature of our "third-person" landscape.

On the one hand, then, is your first-person view, your personal experience of the book of Genesis. On the other hand is the third-person view, the public, true-for-all book of Genesis.

Here's another example. To illustrate these two views for a class, I've sometimes stood my briefcase on end on the desk in front of me and asked if it is the same briefcase for everyone in the room or different. Most have usually said that it was different; after all, what each individual personally sees depends on where they are sitting. Some see mostly the lid or bottom; others see mostly the much narrower sides; in any case, no one sees exactly the same thing. We each have our own first-person view of the briefcase and, indeed, everything.

I then lay the briefcase down and ask if it is laid down for everyone. I have yet to have anyone doubt that it is, and at this point, it's clear that the briefcase is in some sense the same for all. We all see that it is laid down—that it has, in other words, properties that don't change from perspective to perspective. In this respect, there are not a lot of different briefcases on my desk, one for each person; there is only one briefcase laying there, the briefcase as it is from the third-person view, the briefcase that is the same for all.

To show how closely these two views intertwine, I usually go a step further and ask them if it would be the same true-for-all object for an ant that might crawl over it or for a cat. Whatever it might be for such creatures, it's very difficult to imagine that they would see anything like the "briefcase" we saw, and we seem to be back again to the briefcase of the first-person view.

This distinction between first-person and third-person views informs everything we perceive and is of critical importance in our day-to-day lives. If you are crossing a street and don't see a truck coming, your personal first-person view isn't going to keep you alive. What you need is insight into the situation as it is for all regardless of personal view.

The point is that our lives depend on distinguishing between these two perspectives and knowing what is the case from the third-person view. Here's another illustration.

Suppose you and a friend were lost in the desert without water, getting a bit groggy, and you saw an oasis in the distance. Most likely you would ask your friend if he or she saw it too before you decided if it was really there. Why would your companion's opinion make a difference? Because the kind of oasis of interest to you in this situation is the kind that your friend should be able to see too. The kind of water that quenches thirst is the kind that you and I and others can all drink—water that is, in some essential way, the same for all.[5]

In sum, our everyday lives require us to distinguish between the first-person and third-person views. The first-person view gives an object as it is for you or me personally, as we ourselves experience it or think of it. The third-person view gives an object as it is for all irrespective of what you or I might personally see or think.

As I'll discuss shortly, these two views relate to different accounts of reality, while both views combined relate to yet other accounts.

The Two Views before and after René Descartes

As obvious as these two views of reality may be for us today, the distinction in its current form is relatively recent in the history of Western philosophy. Were I teaching before the time of René Descartes (1596–1650 C.E.) and the subsequent eighteenth-century Enlightenment, I doubt that anyone in the class would have thought that the briefcase on my desk was different from one person to the next in the way we do today.[6]

Descartes, who is often considered the principal founder of modern philosophy, gave us a new insight into the distinction between first-person and third-person views, one that much more sharply divided our individual first-person experience from the true-for-all, third-person universe. Especially important, he redrew the line between the two views so as to put literally everything we personally perceive on the side of our first-person views. Per Descartes, a third-person object—a book or briefcase as it is for everyone—is never directly perceived; it is only inferred. What we directly perceive is what our mental faculties create for us to perceive.

Before Descartes, only our experience of something we perceived varied from one individual to the next, not the thing that was presented in our experience. In a commonsense way, the briefcase, sun, or moon that we each personally saw was a briefcase, sun, or moon located outside our bodies and was the same object that others could see (insofar, at least, as our perceptual organs were working properly and capable of showing us what was truly there).[7]

This pre-Descartes worldview is not strange to us today; it remains with us as the worldview of our everyday practical lives coexisting—albeit uncomfortably—with our more modern understanding. In the normal course of my daily activities, I don't think of the chair that I'm sitting on as just my own personal experience of it. I think of it as a chair that others can sit on as well—the chair my wife, for example, sits on when she uses my computer. Likewise, this pen that I'm writing with is a pen that I'm certain others could also write with. The book, the briefcase, the oasis, the sun, and the moon are all things that we perceive and that we assume others can perceive too. Of course, each of us perceives only a tiny fraction of the entire third-person cosmos, and even then, we can make mistakes or be deluded. But as long as our senses are working properly and are not fooled in some fashion, what we perceive seems to be, at least in certain respects, what others perceive. Our first-person view seems to directly reveal at least a portion of the third-person world.

Although we still live our everyday lives as if we directly perceive reality, ever since Descartes it has become very difficult to understand how this can be so.

Descartes wrote at a time when our fledgling modern science was rapidly expanding our understanding of the third-person world.[8] The inventions of the microscope and telescope and the newly central role of mathematics, as well as a host of discoveries about human anatomy, the solar system, and light, revealed a universe very different from the one of our everyday direct experience. On the basis of this and other observations, Descartes concluded that the real world is not something we directly perceive with our sense organs but rather something we know within our minds. As he put it, "Bodies [i.e., material objects, such as a piece of wax held in one's hand] are not, properly speaking, perceived by the senses or by the faculty of imagination, but only by the intellect[;] . . . they are not perceived by being touched or seen, but only insofar as they are expressly understood."[9]

Today, Descartes's conclusion seems inescapable. Every high school science student knows that we do not perceive the world around us except as our mental faculties create it for us. For instance, in the case of visual perception, all that strikes our eyes are patterns of light, which our brains have to interpret before we see anything at all. Indeed, if we saw just what our eyes showed us, we wouldn't see people, cars, houses, or any of the other familiar objects that make up our world. We would see at best only photons traveling at different wavelengths—in other words, just the physical stuff, the matter out of which everything outside our minds seems to be made.

As obvious as Descartes's insight may be, it has left us with an enormous philosophical problem that remains to this day. From the standpoint of physics and biology, no doubt our brains and perceptual apparatus generate the world that we perceive from our first-person views, and we do not directly experience a third-person world that is independent of our minds. Yet from the standpoint of our own first-person views, we *do* seem to directly perceive a third-person world that is independent of our minds. Although the computer on my desk as I personally see it before me can reasonably be only a pattern of neurological signals *inside* my brain, I still see it as *outside* my head and the *same* computer that others also see when they come into this room. Likewise with all the people, books, cars, buildings, and other things that make up the public world we seem to share: our everyday experience of directly perceiving these things seem at odds with what we've learned about our minds and brains and perceptual experience since the time of Descartes and the Enlightenment.

In sum, *both* accounts of reality seem to coexist side by side in our contemporary psyches, and it is far from clear how to reconcile them within one coherent theory.[10]

Descartes's insight also comes with a sharp distinction between mind and matter that continues to shape our understanding of ourselves and the world. Following Descartes, we tend to associate the mind with the realm of first-person experience produced by and interior to our brains and sense organs. Within this realm lies all that is subjective and psychological—our personal perceptions, feelings, values, consciousness, will, sense of self, and the like. External to our minds on the other side of our sense organs is material reality—a third-person view made of atoms, physical bodies, brains, facts, and the rules of causal interaction.[11]

This distinction between inner mind and outer material universe has had enormous implications for religion. The philosophies supporting most contemporary religions originated within the pre-Descartes worldview, and their spiritual landscape is a place in which mind merges with matter—a place where deities, spirits, and even virtue animate our material world and the course of events is governed less by physical laws than by psychological or spiritual ones. Such an interweaving of mind and matter imbues the universe with an organic, living quality. The world we perceive around us is not driven exclusively by the clockwork causality of physical laws but is instead something we can influence through prayer, religious rituals, virtuous behavior, or even just our own force of will.[12]

These days, our post-Descartes worldview makes such a landscape hard to understand. Our disposition to associate all nonmaterial agencies with our minds, with a realm apart from matter, also associates spiritual beings with our minds. Thus God, other deities, angels, and the like are frequently explained as archetypal elements of our psyche (e.g., Jung), social constructs (e.g., Feuerbach, Durkheim), or simply fictions (e.g., Hume, Nietzsche, Freud, Marx).

I'll revisit Descartes later. The new terminology I propose points to a different and, I think, more accurate and helpful way to understand his mind-matter distinction.

The Concept of "Reality"

Along with the first-person and third-person views, another main concept I need to examine before going further is *reality*.

The meaning of the word *real* is closely tied to the two views. The nature of reality wouldn't pose a problem if each of us automatically viewed the world from the third-person perspective—if there were no difference between our personal experience of a thing and what it was for all irrespective of our personal experience. But this is not the case; first-person and third-person views are not the same. Pre-Descartes or post-Descartes, when individuals like you and me look at the world, what we each see depends on our individual circumstances, interests, and perceptual capabilities. We each experience the world from a first-person standpoint and rely on logic, memory, and what others tell us to determine as best we can what might be true from the third-person view—what doesn't change from one individual to another.

Our concept of reality comes from our efforts to understand the third-person view. Historically, the word *real* means "mind independent" or "of the thing itself." This traditional meaning of the word traces to a philosophical debate beginning in the late eleventh century over properties that had seemed since Aristotle to be universals of our natural world (e.g., the species traits that grouped living things into categories such as "horse" or "corn" or "human being"). The question that arose was whether these universals of nature were not instead merely artifacts of our speech or thought. The word eventually invented to refer to what characterized a thing independently of the human mind was *realis*, which became our word "real."[13]

These days, our use of *real* has broadened well beyond its application to just universals. We use the word to contrast our personal ideas of what exists with what in fact ("really") exists, regardless of what that might be. In everyday speech, the *real* contrasts with the conceptual, the imaginary, the fake, the delusional, the appearance (e.g., the real weather in Paris right now versus my idea of what that weather might be). In this respect, the word retains its sense of "mind independent," at least insofar as the mind of an individual is concerned. However, whether "real" refers today to what is independent of the human mind *in general* is another matter. Most of us would call money or political borders real, for example, yet they both depend very much on the human mind.

This question of what is "really" there—what exists from the third-person view—remains to this day as more than just a practical problem of everyday life. In our philosophical efforts to better understand ourselves and the universe we live in, the word *real* points to the problem that Descartes in particular left us of a disconnect between what we perceive and what really exists, a gap between first-person and third-person views that seems contrary to our personal experience of directly perceiving the third-person world.

Take, for instance, the things that we perceive around us in our everyday lives, especially human artifacts such as books and briefcases that are made by and for human beings. Certainly these things are what they are at least in part because of properties that come from the human mind and are not real in the traditional sense of strict mind independence. On the other hand, we certainly think of such things as real in our day-to-day activities. Indeed, if the properties that made something a book or briefcase were not in fact real—if they did not exist in the world outside our personal minds and first-person views—then how would we identify what we saw before us as a book or briefcase? How would we tell it apart from a hat or a teapot or anything else we might imagine?

The result is that many if not all of the things we perceive around us seem to be both mind independent and not mind independent at the same time—both real and not real. Again, we seem both to have and to not have direct perceptual access to reality.

Notice, too, how this particular problem with reality applies only to our post-Descartes worldview. For our everyday, non-Cartesian worldview, the problem disappears because our day-to-day participation in the world comes with a sense of reality as something we directly perceive, something with properties that are not necessarily mind independent. If I tell you, for example, that my desk is plain but functional, I'm telling you about features of my desk that depend at least somewhat on the human mind—features that I think you as a human might recognize but cats and insects would not. Even though such properties are in certain respects mind dependent, I still think of them as real, as properties that belong to the desk itself as it is for both our minds.[14]

The Cartesian gap between first-person and third-person views lies at the heart of many contemporary philosophical problems with understanding reality. It has, for example, an interesting history in science. Two of the principal founders of twentieth-century physics, Albert Einstein and Niels Bohr, famously debated for decades the extent of our access to reality, with many other physicists participating on the periphery.

For Einstein, a good theory of science provided insight into a mind-independent universe: "Einstein liked to use the phrase, 'lifting a corner of the veil.' Scientific theories should lift a corner of the veil that separates us from seeing reality. On the other hand, Bohr . . . was deeply committed to the idea that we can never lift the veil."[15] For Bohr, theories of science didn't show us an independently existing external reality but rather provided us with a means of organizing and interpreting our experience.

Their debate continues today. Those taking Einstein's side are the "realists" (in this philosophical sense). To them, the success of scientific research itself implies that we have access to reality. They point to the ever-increasing predictive power of our theories, a predictive power that depends on theories holding true under all circumstances regardless of first-person view. We see this predictive power in our manufactured goods. Cell phones and vacuum cleaners, cars and eyeglasses behave repeatedly as we expect. We see it in our understanding of natural phenomena—our capacity to predict eclipses or cure diseases.

Opposing the realists are the "instrumentalists," who claim that scientific theories do not show us reality itself, only our experience of it.[16] While realists point to science's successes, instrumentalists point to its limitations and to what the methods of empirical science can't show us (recall the limitations of science mentioned in chapter 1). For instrumentalists, atoms, quarks, and the laws of thermodynamics are useful concepts based on human observation and nothing more; we should not presume they point to things that exist apart from human experience. Supporting the instrumentalist position is the way physicists—realists and instrumentalists alike—generally define a physical "law" in terms of *observers*. The orderly behavior physicists seek is defined specifically in terms of properties (in this case, mathematical relations) that remain the same from one *observation* to the next irrespective of time, place, and certain other circumstances.[17]

From the realist perspective, however, it seems impossible for *observer* here to mean "human observer," since the patterns science uncovers appear to have nothing to do with the human or any other mind. They hold true regardless of what individuals know or believe. Animals know nothing about bacteria and antibodies, yet vaccines work just fine for them. The fact that people once believed that the earth was flat and the center of the universe did not make it true, neither now nor then.

Of course, such remarks are themselves observations made by human beings.

In the end, the Cartesian gap makes it hard to know what in fact our theories of reality refer to. No doubt science bases its theories on what is true for all observers and strives to understand the third-person view; that much seems clear. Where realists and instrumentalists differ is over the question of whether the Cartesian gap can be bridged and our scientific efforts can show us a reality beyond the human mind.

Three Accounts of Reality

Questions about the nature of reality like those raised by the Cartesian gap are questions about the third-person view—what is true for everyone as opposed to just oneself. In this respect, all accounts of reality are accounts of the third-person view, accounts of what is true for all.

Most of the time, our questions about the third-person view are of an everyday variety. What time is it?, for example. Will it rain later today? Will I have enough food for everyone if I double the recipe? But we also have other questions about this view, more difficult and disquieting ones about the nature of the third-person view itself.[18]

I am concerned in this book with the second, philosophical sort of question and how we might make sense of our contradicting answers to these questions. I have already discussed exclusivist, inclusivist, and pluralist alternatives and grouped exclusivism and inclusivism together to distinguish two general strategies for resolving these contradictions. In the discussion to follow, what I call a "third-person" account of reality is one that, like exclusivist and inclusivist alternatives, seeks to understand the third-person world in terms of universal truths that hold for all irrespective of our first-person experience. What I call a "first-person" account is one that, like pluralist varieties, seeks to understand the third-person world in terms of first-person experience and what is actually present therein. Both third-person and first-person strategies give accounts of the third-person view; they both give accounts of reality. The strategies differ in how they explain the third-person view, where they find a rationale for those things of our world that seem to be true for all.

I have also distinguished between realism and instrumentalism in the philosophy of science. While science itself seeks to show us what is universally true for all possible observers and thus gives us a third-person account of reality, our philosophical efforts to understand what science shows us need not give a third-person account. Thus realists trace our theories of science to what is presumed to be a genuinely true-for-all, mind-independent reality, which is a third-person account. Instrumentalists, on the other hand, try to explain the seemingly universal predictive power of science's theories on the basis of our first-person awareness of a world, which is a first-person account.

Besides these two strategies, I'll also discuss philosophical efforts to combine both bases for reality within a single coherent framework. These strategies give what I call "dualist" accounts.

I've structured this book around these three basic ways to understand reality. The next part of the book examines each strategy and how the differences among them relate to differences in religious views as well as to some of our most intractable philosophical problems. The last part of the book explores how the three strategies might complement each other and what that might imply about the universe we live in as well as ourselves as human beings.[19]

Summary

I have distinguished two views of reality: the first-person and third-person views. On the one hand is what you or I personally experience; on the other hand is what is true for all, irrespective of what you or I might personally experience.

Most of the time, our principal problem with the difference between these views is that they don't always coincide. We don't always know what is true for all—what is, as we say, *real*. *Reality*, by which we typically mean the third-person view, is often difficult if not impossible to determine.

Since the seventeenth century, we have faced another and in some ways more troublesome problem with the two views. It's become clear that we do not directly perceive a third-person world that is independent of our minds. There is a gap between our first-person view and whatever might exist outside our minds that makes it very difficult to understand how the two views interrelate. Responses have varied. Some, such as the realists, think that the gap can be bridged, that we do have some sort of first-person access to a third-person universe beyond our minds. Others, such as the instrumentalists, claim that the gap cannot be bridged; our efforts to understand reality are inevitably creations of and limited by our human ways of thinking and perceiving.

Like the distinction between exclusivists and inclusivists on the one hand and pluralists on the other, realists and instrumentalists illustrate contrasting ways to understand reality that are closely associated with the distinction between third-person and first-person views. In the following chapters, what I call a "third-person" account of reality is one that is anchored in what remains the same for everyone irrespective of first-person experience. What I call a "first-person" account is one that is anchored in what is actually present in our individual first-person views. Accounts that attempt to combine both bases for reality into a single coherent framework are what I call a "dualist" account.

The next six chapters of the book further explore these three strategies for understanding reality. The last part of the book proposes a way to understand all three options as complementing each other.

PART TWO

Three Themes

CHAPTER THREE

Universals and Particulars

The Third-Person View Divides into Universals and Particulars

I have said that our first-person and third-person views of the world relate to two conflicting ways to account for reality, while both views combined give yet another approach. None of the options seems to be entirely satisfactory. Each has its strengths but also its weaknesses.

The next six chapters examine each of these three strategies in turn, starting with accounts associated with the third-person view. Hinduism will be my principal illustration of this option. But before I turn to Hinduism, I want to discuss the third-person view in regards to the very old distinction between universals and particulars. It is impossible to understand the concept of God or the Hindu concept of *Brahman* in any philosophically rigorous way without understanding this distinction.

Consider again the briefcase example of the two views from the prior chapter. My illustration distinguished between the briefcase as we personally see it (the first-person view) and the briefcase that is there for all (the third-person view).

Now consider the example of the book of Genesis. I used it to illustrate the same distinction, but you might have noticed that, unlike my briefcase, the third-person book of Genesis referred to more than a single physical object.

First, there is the particular book of Genesis you are holding and reading right now—the book that you could lend to a friend to read. In respect to it being a particular third-person object like this, the book of Genesis is like the briefcase.

But of course your friend doesn't actually have to borrow the book in your hands in order to read it; many other copies are also available. Genesis's third-person existence includes not simply the particular book you might hold and read insofar as others might also hold and read it; it is also a book that has been reprinted innumerable times and comes in a variety of shapes and sizes.

Second, then, is the book of Genesis that spans all particular instances and remains the same from one copy of the book to the next. Big print or small print, handwritten or printed, it is still in some sense the same book of Genesis.

The third-person view thus encompasses two categories of things. First are *particulars*, the book of Genesis you are currently holding. Second are *universals*, what stays the same for or can be said of a number of particulars: in this case, the book that remains the same from one particular copy of it to the next. Both are aspects of the third-person view; both remain the same from one observer to the next.

For another example of the distinction, consider a word such as *red*. Notice how it can be printed in a variety of different ways: red, RED, *Red*, *RED*. Regardless of how different each particular example is, however, all are recognizably the same word. Something about each printed instance of the word is universal; it remains the same and permits recognition. (In philosophy, the universal "red" that remains the same from instance to instance is the *type*; the instances of the word are the *tokens*.[1])

In this distinction, particulars need not be clearly defined objects such as a book or a person; they can also be masses or collections of things. Standing on the shore of the Atlantic Ocean, I'd likely identify what I saw before me as an instance of "water." Even though the water I see is not well delineated at all, it is still *this particular* water and not some other.

Also in this distinction, universals need not be what is the same from one moment to the next. They can also refer to changes in our world when one state of affairs characterized by a certain set of properties is always followed by another state of affairs characterized by a different set of properties. Thus when hydrogen and oxygen are combined and heated sufficiently at normal atmospheric pressure, they predictably combust and produce water vapor. To the extent that the behavior of our world can be reduced to such rules, these rules are also universals.[2]

Like the distinction between first-person and third-person views, the distinction between universal and particular pervades and is essential to all human life. To look at things and recognize them is to see particulars as instances of universals. You and I couldn't distinguish prey from predator, friend from foe, or even food from the plate it sat on were we not able to place things into categories—in other words, to see them as instances of universals.

It is hard to think of anything in human life that doesn't involve this distinction. All the properties shared by the things making up our world are universals, as are all numbers, words, rules of logic and math, and laws of physics.[3] The individual occurrences of properties, numbers, words, and the rest, on the other hand, are particulars.

Similarly, the logic and structure of DNA are universals instanced in all DNA molecules. The zoological classes into which we classify creatures are universals instanced in the individual creatures falling into these classes. Every word we speak is a universal given a particular meaning by its present context.

While the existence and importance of this distinction in our day-to-day lives may be obvious, the nature of universals and their relation to particulars are not and have been highly controversial in both Western and Eastern philosophy for a very long time. As noted in the previous chapter, our word *real* traces back to a debate beginning in the late eleventh century over whether universals are independent of the human mind, a debate that continues today in various forms. Thus instrumentalists disagree with realists over whether theories of physics refer to mind-independent universals. Thus certain properties that describe, say, a particular house or boat are universals that seem to both come and not come from the human mind. The nature of universals and their relationship to the particulars of everyday experience have also been debated extensively in the East, such as in the disagreement between Hindus and Buddhists over the nature of reality that I will talk about later.[4]

Some Historical Roots of the Universal-Particular Distinction

The distinction between universals and particulars is at least as old as language, since the meanings of words are universals insofar as they apply to more than one particular thing (i.e., insofar as words are not proper names like, e.g., *Plato* or *Mount Everest*). Moreover, every spoken word is itself an instance (i.e., "token") of a universal (i.e., "type") in the same way that a printed word is. The uttering of *Plato* in one moment is an instance of all utterings of the word.

The distinction also shows up in ancient religions. Some deities were particular things—the sun, moon, certain mountains, or cities (to the extent that they were deified). But other deities were universals insofar as they stood for classes of things—gods of sheep or corn or fire, for example. In these cases, a god typically captured the idea of the sheep or corn or fire as a generalized agency of nature covering all things of a certain type. The deity was that which accounted for, among other things, why *all* sheep or corn or fire behaved everywhere in much the same way.

One of the ways we know that people thought of deities this way is because the names of deities were often the same as the words used to name the objects they personified. Thus in Mesopotamia, *Utu* referred to both the sun that people saw in the sky and the sun personified as a deity. Likewise, *Ezen* meant both grain and the grain god; *Imdugud* meant both rain cloud and the rain cloud god. In India, *Surya* referred to both the sun and the sun god; *Agni* to fire and the god of fire; *Vāyu* to wind and the god of wind. Similarly for other aspects of human life: in India, *Kāma* meant both desire and the god of desire.[5]

Personifying a class of things as a deity emphasizes the importance of the universal as a thing in itself. To seek to understand and relate to such a deity is to look at corn or barley or fire as a class of things. And indeed, it is the behavior of the class, not the individual by itself, that must be understood if one is to grow crops and build fires.

The development of philosophy in the first millennium B.C.E. transformed and formalized the concepts of universal and particular.[6] According to tradition, Western philosophy began in 585 B.C.E. In that year, Thales of the town of Miletus on the southwestern coast of what is now Turkey is said to have predicted an eclipse of the sun. For Thales, such an eclipse was not the capricious activity of deities but rather something explainable in terms of underlying natural principles. Thales and his successors looked for ultimate reality not in a diversity of gods and goddesses but in universal rules that existed throughout all things.

It is now commonplace to think of the universe as encompassing all things linked together into a whole, something that has its own nature and that expresses itself through certain universal properties such as space, time, mass, and energy. But this idea of the universe as one thing was not a part of the Mediterranean worldview before the time of Thales. Although the Mesopotamians had a concept of a primal chaos out of which the deities formed our heavenly and earthly circumstances, they didn't have a concept like "universe" that referred to everything as *itself* a thing. Heraclitus (c. 540–480 B.C.E.) was the first we know to have used the Greek word *kosmos* (from which we get "cosmos" and "cosmology"), although the philosophers before him no doubt had some similar idea in mind when they spoke of all things as governed by universal principles.[7]

This new conception of the cosmos as governed by universal properties was not unique to the Mediterranean. Around this time in China, scholars had started to talk about the Dao, the "way" of all things. In India, Hindus were fleshing out their already existing idea of Brahman as the one eternal and universal basis of all things, while the Buddha challenged Hindu orthodoxy with his idea of the cosmos as a process of change.[8]

For the early Greeks, the nature of this new universe raised many questions associated with the difference between particulars and universals. Heraclitus (c. 535–480 B.C.E.) and Parmenides (sixth or fifth century B.C.E.), for example, disagreed over whether existence was ever-changing or eternal. For Heraclitus, every experience was fundamentally new and different (he reputedly said that "it is not possible to step into the same river twice"[9]), which points to nature as a process of change. Such an account of nature finds its support in particulars in the sense that particulars differ from one time and place to the next. Parmenides, on the other hand, insisted that all existing things had their existence (their "is"-ness) in common and that existence or being was of one kind, unchanging, and eternal. In other words, he emphasized universals, or what stays the same from one time and place to the next.

The two Greek philosophers who contributed the most to our understanding of universals and particulars were Plato and Aristotle.

One of Plato's (427–347 B.C.E.) most important and far-reaching distinctions was between sense experience (*aisthēsis*) and intelligence (*nous, noēsis*): through sense experience we grasp particulars; through intelligence we grasp universals. Our first-person sense experience shows us just what is currently here and now, not what extends over many here and nows. If, for example, we examine a chair as we directly perceive it—list all its properties as carefully as we can—we invariably find ways it differs from other chairs. Some chairs have four legs, some three, others only one; some have armrests, others don't; some are bean bags; some are built for children. While *chair* as a universal covers all these variations, the chair that we directly see and touch does not. When we look at a chair, the book of Genesis, a printed word, or anything else, what we see before us is *this* chair here, not all chairs; *this* version of Genesis, not all versions of Genesis; *this* printed instance of the word *red*, not all printed instances of the word. In order to grasp the universal *chair* or *Genesis* or *red* and recognize what we see as an instance of a universal requires another faculty of awareness other than sense experience. It requires the capacity to correlate various here-and-now experiences and determine what is the same and different among them—a capacity or faculty we typically call intelligence, reason, or cognition.

Plato often uses mathematics, especially geometry, to illustrate how we grasp universals differently from particulars. The geometry of a right triangle, for example, holds for everyone, but nowhere do we actually see a perfect right triangle. Similarly with the ideas of "one" or "two" or "equality" as in "one plus one equals two": these ideas also don't seem to exist in the world of our sense experience. Holding up two sticks of equal length side by side doesn't show us the universal idea of "equality," just a situation in which that idea might be applied.[10]

This observation that sense experience by itself doesn't show us universals has been echoed in various ways by many philosophers.[11] It plays an important role in philosophy and will reappear frequently in later chapters of this book.

Aristotle, like his mentor, Plato, bases what is arguably his most fundamental distinction on the difference between particulars and universals. What makes things what they are is their "substance" (*ousia*), which he divides into two kinds: "primary substances" (or just "substances") and "secondary substances," or "universals." Primary substances pertain to particular things; they make each particular what it is and numerically just one thing. The substance of a person such as Socrates, for example, is what makes Socrates "Socrates" and not someone or something else. On the other hand, secondary substances (universals) are what makes one thing like another, or what remains the same (numerically singular) over a number of primary substances. The property of "humanness"—a secondary substance or universal—applies not only to Socrates but to other people as well. For Aristotle, a primary substance is what is characterized by universals without itself being a universal, while a universal can apply to many primary substances.[12]

Both Plato and Aristotle were enormously important in shaping our modern view of the world. Much of subsequent Western philosophy can be thought of as efforts to work out the details of what Aristotle and Plato observed. And of these details, perhaps the most fundamental and vexing concern the nature of universals and their relationship to the particulars of our perceptual experience.

Universals and Particulars as Principles of Nature

Because of the importance of the universal-particular distinction in philosophy and everyday life, I want to provide an extended illustration of these two concepts as principles of nature.

Imagine yourself taking a Physics 101 course for the nonscientist. You and your classmates open the textbook to the chapter on atoms and begin by reviewing what you already learned in high school. You read how all atoms have a nucleus made of protons and almost always also neutrons and how all protons and neutrons are themselves made of quarks. You look at diagrams and read how electrons always configure themselves in fuzzy, concentric shells around the nucleus and how each shell never has more than a certain maximum number of electrons. You look at actual pictures of atoms taken with electron microscopes or using other techniques. You also read how the number of electrons in the outermost shell determines how atoms behave chemically and how these rules of behavior account for all the elements and compounds out of which everything we see around us is made.

It is not, however, the details of this description that are of most interest here but rather the nature of the description itself. Even though your class is studying

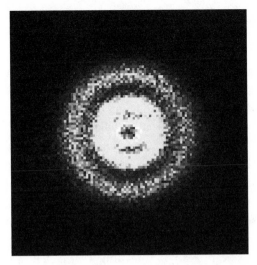

FIGURE 1. A particular atom. The orbital structure of the electron of an actual hydrogen atom, captured using photoionization microscopy. Photograph © 2013 American Physical Society. Courtesy of Dr. Marc Vrakking.

$2n^2$ = Maximum electrons per shell

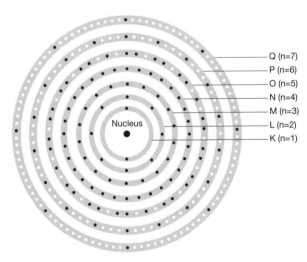

Q (n=7)
P (n=6)
O (n=5)
N (n=4)
M (n=3)
L (n=2)
K (n=1)

Nucleus

FIGURE 2. Atoms in general. Schema for the maximum number of electrons in each shell, with black dots representing the maximum up through ununoctium, the heaviest atom we've so far found or created. Drawing by CranCentral Graphics.

atoms, the textbook certainly does not describe each individual atom in the universe. It describes instead what all of them have in common—their underlying universal nature. In other words, the textbook is not so much about particular atoms themselves as it is about *the* atom—what might be called the "generic," "paradigmatic," or "archetypal" atom. This single atom blueprint—let's call it the "archetypal" atom—embodies the orderliness of all atoms; their common, fundamental nature or essence; the universal rules that all atoms follow.

Many people—for example, the instrumentalists I talked about earlier—would insist that this archetypal atom doesn't exist except in the minds of scientists; it's merely a useful idea, a conceptual convenience for organizing our experience. Regardless of whether you fall on the realist or the instrumentalist side of this debate, however, these rules of nature enter into our experience as universals, as what holds true throughout the world as we know it. Perhaps the concept of an atom archetype doesn't refer to something that is real in the sense of being entirely mind independent; nevertheless, it points to a universal aspect of at least human experience. More than that, the archetypal atom enters into our experience as a basic constituent of the third-person view, a reality underlying *all* things, including the living creatures and inert matter we experience as other than ourselves. Mind independent or not, this universal atom archetype (or whatever you prefer to call it) presents itself to us as something very different than atom instances.

There are six differences between this archetypal atom and its instances that I'd like to point out.

First, they differ in *number*. There is one archetypal atom, while there are many atom instances. In the Physics 101 textbook, all atoms share only one fundamental nature. It is one and the same for every atom instance. Atom instances, on the other hand, are many.

Second, they differ in *location*. The archetypal atom exists throughout space and time wherever (at least) there are atoms, while each atom instance exists only at one particular space-time location.[13] Additionally, the entire archetypal atom with all its properties exists at each location, not one piece here and another piece there. Indeed, the archetype is made not of spatiotemporal parts such that one part could be here and another there but rather of properties, rules of behavior, and the like (next difference).

Third, they differ in *composition*. The archetype is abstract or incorporeal—a set of rules or properties that describe the behavior of all atoms. An atom instance, on the other hand is corporeal; it is made of parts that exist in space and time, with the nucleus at one place and the shells ranged around in other places. (See text box on pages 36–37.)

Fourth, they differ in *mutability and life-span*. The archetypal atom is unchanging and lasts forever (or at least as long as there are atoms), while individual atom instances have constantly moving parts, can be affected by circumstances, and have finite life-spans (though for atoms that apparently can be a *very* long time). We presume that the same rules of atom behavior have existed since the beginning of the universe, while individual atoms are subject to change and may be created or destroyed.

Fifth, they differ in *how they are known*. The archetypal atom can't be directly observed. It can only be inferred from our experience with many atom instances. Atom instances, on the other hand, can be empirically observed. They interact with each other and can be seen with the aid of electron microscopes, for example. In general, as Plato said, sense experience shows us not universals or types or kinds of things but rather instances of them as they appear in the cause-and-effect dynamics of this or that situation. The red color or the human being that we see before us is this particular color red or that particular human being. The redness of all red things or the humanness that all humans share is an inference based on a number of such experiences.[14]

Last, they differ in *causal role*. The archetypal atom might be described as "law-giving" in the sense that it determines behavior, while atom instances are "law following." The archetype embodies all the rules or principles that govern the behavior of atoms; the individual atoms instantiate the rules. Note that these principles are not just descriptive; they are also *prescriptive*. To know the archetype is to know within certain limits what an atom *will* do.[15] It is because atoms behave in these logical, predictable ways that we are able to make things out of them that also behave in predictable ways—cars, televisions, cell phones, computers, digital cameras, or any other manufactured product.

This last difference expresses itself in another way as well. The kind of causality that relates atom instances to the archetype is not the same as the kind that relates atom instances to each other. Atom instances relate to one another in the interactive, cause-and-effect fashion that we normally associate with the idea of causation. An individual atom is not "lawgiving" in its effect on other atoms. Rather, it collides with or bonds with or weakly interacts with other atoms. One thing happens, which causes the next thing to happen, which causes the next, and so on—a little like an endless cascade of dominoes wherein the occurrences of each moment give rise to the next.

Thus atom instances participate in two kinds of causality. On the one hand, in their relationship to the archetype, individual atoms follow rules; their behavior is predictably the same from one instance to the next. On the other hand, in their relationship to each other, each atom plays its own individual, interactive part; each atom's behavior varies in response to changing causal conditions.

THE DIFFERENT COMPOSITIONS OF
PARTICULARS AND UNIVERSALS

One of the most obvious and important differences between particulars and universals is their composition. Particulars are "corporeal"
in the sense that they have "part-whole construction"; universals
obey "class logic."

To have part-whole construction is to be made out of parts
that are, in turn, made out of other parts. A particular oak tree
is composed of roots, trunk, limbs, leaves, acorns—all distributed spatially in a characteristic way. A particular leaf or acorn
has its own spatially arranged parts, each of which is made out of
plant cells, all of which are themselves particulars and made out
of other parts.[16]

Universals, however, do not have part-whole construction
like this; they define groups of instances that obey class logic. For
example, the property of "being an oak tree" collects all oak trees,
real and imagined, into the same group. The property marks out
a category or class of things whose members all have that property. Class logic has a hierarchical structure somewhat like part-
whole construction. Going up the hierarchy, oaks are a type of tree,
which is a type of plant. Going down the hierarchy, the class of all
oak trees divides into white oaks, bur oaks, pin oaks, and other
varieties.

Members of a class can be particulars—a pin oak in one's backyard, for example. But while particulars can participate in classes,
they don't themselves form classes of things. If a particular oak tree is
a class at all, it is a class of just one—a class whose only member
is itself.[17]

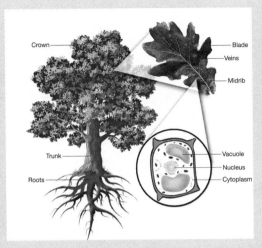

FIGURE 3. Part-whole composition of an oak tree. Drawing by CranCentral Graphics.

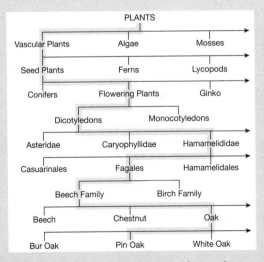

FIGURE 4. Class-logic structure of an oak tree. Drawing by CranCentral Graphics.

Atom instances may follow rules, but how the rules are expressed differs from one instance to the next.[18]

In sum, the archetypal atom contrasts with atom instances in the following ways:

1. *Number.* Archetype: one. Instances: many.
2. *Location.* Archetype: omnipresent. Instances: specific spatiotemporal presences.
3. *Composition.* Archetype: incorporeal. Instances: corporeal (made of parts).
4. *Mutability and Life-Span.* Archetype: unchanging/eternal. Instances: changing/finite.
5. *How Known.* Archetype: inferred only. Instances: empirically observable, at least in principle.
6. *Causal Role.* Archetype: lawgiving. Instances: law following in regards to archetypes; causally interactive in regards to each other.

It should be obvious by now how similar these qualities of an archetype are to many of the properties that characterize the Western idea of God. Western monotheism typically distinguishes God from creation using roughly these same traits. God is one, not many; omnipresent, not localized; incorporeal, not corporeal; unchanging and eternal, not changing or finite; inferred (or otherwise nonsensuously revealed), not empirically observed; and lawgiving, not law following.

We are still far from a complete description of monotheism's idea of a supreme deity. There is no sense here of God as an aware being of a kind who might respond to prayer and ritual. Nevertheless, the idea of God as the fundamental nature of the universe does make sense out of most of the deity's other main characteristics. It easily explains, for instance, how God might be an incorporeal cause of the way events unfold and how God is everywhere in time and space at the same moment. Also, as we will see shortly, it helps explain the Hindu concept of *Brahman.*

The Ambiguous Relationship of Universals and Particulars

I want to point out a certain peculiarity about this distinction between universals and particulars that has puzzled many people for a long time.

Notice how a universal both *is* and *is not* its instances. On the one hand, the "archetypal atom" is not the same as the individual atom instances we try to glimpse with electron microscopes; the two differ in the various ways I described. Likewise, the type of creature we are—"human being"—is not itself a particular person; the class of all pencils is not itself a particular pencil such as the one in my hand. On the other hand, notice how each instance of a universal is just that: an *instance of a universal*. Every atom *is* an atom; every human being *is* a human being; every pencil *is* a pencil. The particular things making up our world seem in some sense to *be* the universals they instantiate.

Not surprisingly, this peculiarity has caused difficulties from the beginning of philosophy. Plato and Aristotle, for example, disagreed over where universals resided. Plato claimed that universals were separate from particulars and resided in a special realm of their own, while Aristotle argued that they were in the particulars themselves. Aristotle's view is more consistent with our everyday experience of objects as themselves having properties and is, I expect, much more common today, although the abstract, universal nature of math and logic remains a compelling argument for Plato's view.

Notice, too, that we run into a similar problem when we approach the universal-particular relationship from the standpoint of particulars and the properties that describe them. Just as a universal has an ambiguous relationship to its instantiations in particulars, a particular has an ambiguous relationship to the properties that identify it as being similar in various respects to other particulars. On the one hand, surely the properties that identify certain objects as a "person" or "atom" or "pencil" can be distinguished from the particular people, atoms, or pencils that the properties identify. Insofar as they are not proper names, such properties are universals, and they differ from particulars in the ways mentioned. On the other hand, it's proven very difficult to understand what a particular thing is apart from these identifying properties. What would in fact be left of a particular person, atom, or pencil were we able to strip away all the universals that characterize it and expose just the "bare particular," as it's sometimes called? Would anything be left at all?[19]

Universals, Particulars, and Accounts of Reality

Questions about the nature of universals and their relationship to particulars have long been central to debates over the nature of reality. With philosophy founded in part on the idea that the third-person world is a single thing governed by universal first principles, most efforts to understand the nature of reality have presumed the existence of these third-person universals and have looked to them to explain the particulars of our first-person experience. Plato's forms and Aristotle's substances illustrate this strategy for understanding reality. Thales's prediction of an eclipse depended on such a presumption about the nature of the cosmos. Such accounts of reality are what I'm calling third-person accounts. They seek to understand the third-person world in terms of universal truths that hold for all irrespective of our first-person experience.

Unlike third-person accounts of reality, unalloyed first-person accounts are harder to find in the West until Catholic scholars in the late Middle Ages started to wonder if universals might depend on the human mind. Much earlier, Greek Sophists and Skeptics like Gorgias and Pyrrho criticized accounts of reality based on universals of nature but offered little more than that in the way of a first-person account. Of the early Greeks, perhaps Heraclitus came closest to a thoroughgoing first-person account. Although he described nature as animated by a first principle, an exchange of heat or "fire," he also described it as a process of change in which the present is always unstable, ever-changing, and new.[20] In this respect, he gave priority to what an individual directly experiences (particulars) over what is true for all irrespective of an individual's direct experience (universals).

This debate over the role of universals in explaining the world we perceive is the first of two principal debates that shaped our Western concept of reality, neither of which has ever been satisfactorily resolved. This first debate concerns universals alone, and as discussed in chapter 2, our word *real* dates from this debate's reemergence in the late Middle Ages. In answer to the question of where universals come from, realists claim that they come from nature and tell us about the universe as it truly is. Conversely, antirealists claim that they are inventions of our minds and that the particulars of our first-person experience are all we might know of the universe we live in. (Among antirealists, "nominalists" trace universals to words and "conceptualists" trace them to concepts.) Insofar as these positions are carried through into explanations of our third-person view, realists give what I'm calling third-person accounts, while antirealists give what I'm calling first-person accounts.

The second principal debate shaping our concept of reality was precipitated by Descartes's distinction between interior mind and exterior matter and his insight that the world we perceive around us is created by the mind of the perceiver. The main issue then became the extent to which not just universals but everything—the entire third-person universe—had its origin in the mind. In this second debate, realists claim that the third-person universe is something that exists independently of the mind, while antirealists claim that it is not.[21]

In this second debate, the realist and antirealist positions do *not* provide a straightforward way to distinguish between third-person and first-person accounts of reality. While all realists give third-person accounts, many antirealists also give third-person accounts. "Idealists" (in the philosophical sense of the term) such as Leibniz and Berkeley resolve Descartes's mind-matter dualism by tracing the true-for-all nature of the third-person world to universals of the mind, which is a *third-person* strategy, not a first-person one. They presume the existence of universal truths that hold for all irrespective of our first-person experience, except that, in this case, these universals come from the mind as itself in some respect the underlying, universal basis of our cosmos. Conversely, first-person strategies emphasize first-person experience and particulars, as they did before Descartes. Here universals are conceived of as derivative—generated, for instance, by the participation of individuals in social groups. This is the pluralist approach and the approach of instrumentalism in the philosophy of science. This is also the general approach of postmodern/poststructural/existential philosophies—those philosophies that loosely comprise "continental philosophy."

The variety of accounts that in some way trace reality to our minds raises a question of importance to later chapters. I've associated the third-person view and reality with what is true for all, but must the third-person view that interests us in our accounts of reality always be that which is true for absolutely everyone and everything? Is this in fact how we (or our mental faculties) distinguish the third-person view from the first-person view in our everyday lives?

These questions bring us to a somewhat technical matter, an ambiguity in our grammatical first-person plural. The grammatical first-person plural ("we") has two senses. One is "collective"; it applies to everyone together. The other is "distributive"; it applies to everyone individually. For example, if I said, "There were four of us at the table yesterday, and we had all written a book," I could mean that we collectively (i.e., together) wrote a book or that we distributively (i.e., each) wrote a book. In the first collective case, we wrote a total of one book; in the second distributive case, we wrote four.[22]

When considering the difference between the third-person and the first-person views in philosophy, clearly the distributive sense of "we" pertains to the first-person view; it refers to us as a group but regarded as individuals. But what about the collective sense of "we"? If I share a culture and a language with a group of people such that we share collective objects in common (e.g., money, word meanings, political borders), then might not this collective sense of the grammatical first-person plural refer to a *third-person view* rather than a first-person view, at least in certain cases? It may not be the true-for-absolutely-everyone third-person view, but might it not still be the third-person view of primary interest to this subset of individuals? (I'll return to this subject of group-based third-person views in chapter 5.)

Despite the different realist versus antirealist debates and the ambiguity of our grammatical first-person plural, the central issue addressed here can be stated fairly simply: How do we account for the true-for-all world we perceive around us? Do we find our explanation in what is in itself true for all, whether it be an external universe or universals associated with the mind? Do we find it, instead, in what is actually present in first-person experience and interaction, with universals generated, for example, through habits of thought or through rules or shared ways of thinking we invent? Or, again, do we find it in some combination of both approaches?

Summary

The third-person view divides into particulars and universals. On the one hand are particulars such as the book or briefcase that we might be looking at; on the other hand are universals such as the properties that identify all books or all briefcases. Universals (properties, relations, categories, etc.) are what is common to a number of particulars; particulars instantiate universals and have no instances besides themselves.

Of the two, our first-person sense experience shows us only particulars: *this* color red on the cover of my book or *that* person sitting over there, not the "red" of all red things or the "humanness" of all human beings. To know universals, we have to correlate different first-person perceptions and find what does and does not stay the same.

Eastern and Western philosophy grew out of the idea that the universe as a whole was one thing with certain properties that held throughout. Accounts of reality anchored in such universal properties or principles are what I call "third-person" accounts. Philosophies that argue against this strategy for understanding reality and seek instead to base their accounts on particulars and what is actually present in first-person experience are what I call "first-person" accounts of reality.

Many of the properties we typically attribute to God are those that distinguish a universal from its instances. Thus God is typically distinguished from the everyday world as being one, not many; omnipresent, not localized; incorporeal, not corporeal; unchanging and eternal, not changing or finite; inferred (or otherwise nonsensuously revealed), not empirically observed; and lawgiving, not law following.

Universals and particulars have an ambiguous relationship, which complicates our understanding of them. A further complication is the Cartesian gap between interior mind and exterior matter, which raises the question of whether universals, particulars, and possibly all of reality might come from the mind.

Despite these and other complications, the central question of interest here is fairly simply stated: Where do we find our rationale for the third-person view? Do we find it in what is in itself true for all, be it a mind-independent universe or universals associated with the mind? Do we find it in what is present in first-person experience? Or do we find it in both?

CHAPTER FOUR

Hinduism and the Third-Person View

The same stream of life that runs through my veins night and day runs through the world and dances in rhythmic measures. It is the same life that shoots in joy through the dust of the earth in numberless blades of grass and breaks into tumultuous waves of leaves and flowers. It is the same life that is rocked in the ocean-cradle of birth and of death, in ebb and in flow. I feel my limbs are made glorious by the touch of this world of life. And my pride is from the life-throb of ages dancing in my blood this moment.

From Gitanjali, "Song Offerings" (verse 69) by Rabindranath Tagore
(1861–1941), Indian writer in many genres and Asia's first Nobel Laureate

Why Hinduism?

What I call a third-person account of reality is one that seeks to understand the third-person view in terms of the third-person view itself—what holds true for all irrespective of first-person experience. It contrasts with a first-person account in which one seeks to understand the third-person view in terms of what is actually present in first-person experience.

I've chosen Hinduism to illustrate a third-person account of reality because among the major living religions, it offers one of the oldest and most thoroughly developed and highly regarded versions of this strategy. Some of its many scriptures were written down in Sanskrit[1] over three thousand years ago, and certain of these scriptures characterize God—*Brahman*, as the Hindus say—as the universal basis of all that exists. More explicitly than perhaps any other major religion, these texts teach that Brahman is not a being apart from existence but rather the fundamental nature of all that is.

That said, it would be unfair to Hinduism to paint it with too broad a brush. Like the other religions I will use to illustrate philosophical strategies, Hinduism has many philosophical schools; it encompasses many different views of reality.

My following account of Hinduism focuses principally on its "nondual" school (*Advaita-Vedānta*) and those passages from Hindu scripture most often used to support this view. This school has been the principal face of Hinduism to many Westerners and an enduring source of spiritual inspiration for a great many people worldwide.[2] I will, however, also touch on some of Hinduism's other main schools, all of which anchor their accounts of reality in third-person universals to one degree or another (i.e., they give what I call either a third-person or dualist account).

Before continuing, let me interject a comment about the structure of this book. This chapter and the chapters on Buddhism and Western theism are longer than the other chapters. These chapters illustrate the three accounts of reality I'm interested in, and I want to flesh out their contrasting views in more detail than shorter chapters would allow.

Brahman

The Hindu scriptures trace back to four collections of a very old oral tradition written down in Sanskrit beginning about 1200 B.C.E. These collections together are called the *Vedas* ("knowledge"). Numerous enough to fill a good-size bookshelf, they contain mostly hymns to gods, ritual incantations, and magical formulas.

The role of gods and goddesses in accounts of reality is of less interest to my discussion now than it will be later on when I turn to dualist philosophical strategies. Nevertheless, the worship of deities continues to play an important role in Hinduism, and besides the deities mentioned in the previous chapter, I should perhaps mention three more especially important ones that together capture for Hindus the principal dynamic of nature. *Brahma*, *Vishnu*, and *Shiva* are, respectively, the gods of creation, preservation, and destruction—as, for example, in the birth, growth, and death of living beings. These three gods also have different meanings in different contexts and for different people. Brahma, for example, can stand for the creator of the Vedas and Shiva for the destroyer of ignorance.

Around 800 B.C.E., perhaps two centuries before the beginning of Greek philosophy, the Vedic period ended, and India began a period of philosophical speculation collected in texts called the *Upanishads* (*upa*, meaning "supplementary"; *ni-shad*, meaning "to sit down near" a spiritual teacher). For six hundred years, more texts were added to the Upanishads, which became part of the Vedas. Around 200 B.C.E., the period ended when interest shifted from philosophical writings for the priestly elite to writings more suitable to the general population, such as religious histories and tales (the *Rāmāyaṇa*, *Mahābhārata*, and *Purāṇas*).

With the Upanishads came the elaboration of an idea found occasionally in the early Vedas that the many gods of the Hindu pantheon were but different manifestations of a single ultimate reality.[3] Instead of the hymns to deities and magical formulas of the older scriptures, these new writings are about this one ultimate reality of all things, *Brahman*, as well as how to arrive at a firsthand experience of this reality.[4]

While there are many similarities between Brahman and the God we typically associate with Western monotheism, there are marked differences as well. Most important, Brahman is not something apart from this universe but rather its underlying nature—that which our universe and everything in it *is* at its most fundamental level. Similarly, Brahman is not a being separate from oneself that one might somehow relate to in the way one person might relate to another. Instead, Brahman relates to the universe and to individual people in the way the atom archetype of the prior chapter relates to atom instances.

Indian stories about the creation of the universe are particularly telling in this regard. They are significantly different from what one finds in Judaism and Christianity. In the Judeo-Christian version, the book of Genesis characterizes God as similar to an artisan creating a work of art that is distinct from the artist: "God said, 'Let there be light'; and there was light. God saw that the light was good, and God separated the light from the darkness. God called the light Day, and the darkness He called Night. . . . And [on the sixth day after creating all other things] God created man in His image, in the image of God He created him; male and female He created them" (Gen. 1:3–4, 27).[5]

The Indian versions, on the other hand, usually tell of a being that evolves into and, at least in part, *becomes* what is created. One version of the Hindu creation myth from the early Vedas tells of a "cosmic egg" that hatches and itself becomes the universe. Another version from the Upanishads tells of a solitary Self that existed prior to creation: "However, [this solitary Self] . . . lacked delight . . . and desired a second. . . . This Self then divided itself in two parts; and with that, there were a master and a mistress. . . . The male embraced the female, and from that the human race arose."[6] The pair then divided and mated again, creating cattle, and then they repeated this cycle over and over until all creatures were created.

The analogy of Brahman to the atom archetype goes only so far, however. In the above and other passages in the Upanishads, Brahman is presented more as an aware being than a blueprint or collection of rules describing the behavior of the universe.

In order to understand the Hindu concept of Brahman (and, in later chapters, the monotheist concept of God), two related points must be kept in mind. First, the universe of interest in Hinduism is specifically the universe we live in and experience around ourselves, which is not a universe cut off from us by a Cartesian gap between interior mind and exterior matter. In Hindu philosophy, there is no Cartesian mind-matter divide. The mind is not an internal realm of experience distinct from reality but rather a faculty that, among other things, directly perceives and participates in the particulars of our day-to-day world. As discussed before, this direct experience of the third-person view is our own typical everyday experience today. The sun, trees, cars, and people I personally see are things I assume others can also see.[7]

The second point follows from the first. With the first-person view directly aware of reality to one degree or another, Upanishad philosophy does not presume that what we perceive can be understood apart from the awareness that perceives it. In these scriptures, it is not simply objects of awareness—the people, trees, and so on we perceive around us—that need to be accounted for. Even more important is the knowing subject and the way it is aware. The quest to discover the fundamental nature of all things is a quest to be pursued not in objects apart from the observer but rather in *subject and object combined*—in the union of what some Hindus call the "triple thread" of knower, known, and knowing.[8]

Thus Brahman relates to the world not as an archetype of an inert object like an atom but rather as the archetype of the *aware self*.

Let me reiterate this last point; it is key to understanding not only Hinduism but many other Eastern and Western philosophies and religious traditions. Hindus argue—as do instrumentalists in the philosophy of science and many others—that the universe we are aware of around ourselves is not something to be understood apart from our awareness of it. The concept of Brahman in Hinduism points to this fundamental I-am-aware that many people claim underlies everything we might know about ourselves and the world. Hinduism differs from first-person accounts of reality such as instrumentalism, however, in that this fundamental I-am-aware is a *universal*, a first principle of the cosmos we live in. In these Hindu teachings, everything that exists emerges from a single aware self, and there is no reality nor individual aware selves apart from this aware self principle.

Hindu scriptures flesh out Brahman's nature as this archetypal aware self in a variety of ways. Often Brahman is referred to as simply the Self—*Ātman* in Sanskrit. In Hinduism, ātman is the word for a person's most basic underlying being. Thus "Ātman" in this capitalized sense refers to the universal Self of which all particular selves are instances, the Self that is everybody's self. (Recall in this regard how the names of deities often do double duty as both a personified universal and the instantiation of that universal; for example, *Agni*, or fire, refers to fire as a deity as well as to fire as we experience it.)

The concept of Brahman may refer to the self as a universal, the Ātman; however, this archetypal self is not a universal as we might usually think of it. The concept of Brahman points (as best a concept can) *beyond* the universal-particular distinction to something more fundamental within which both concepts are subsumed.[9] Brahman is clearly a universal, but Brahman is also the source of particularity and difference, that which explains how particulars come to be individuated, each with unique properties. Brahman is not a mere blueprint or class of instances; this basis for all is the *single* aware self out of which arise *all the different* aware selves and the other particulars that we see around us.[10]

Significant in this regard is how Hindu scriptures describe the awareness associated with the Ātman. Hinduism's most respected and popular text, the *Bhagavad-gītā*, describes Brahman as that which sees through all eyes, hears through all ears, grasps with all hands, and walks on all feet.[11] In the Upanishads, one often-cited passage identifies Brahman with the different levels of consciousness one finds in sleep and waking. It begins by speaking of Brahman as the underlying universal Self, Ātman, and then describes this Ātman as that which is conscious of external objects in the waking state, that which is conscious of internal objects in the dream state, that which is conscious of no objects in the deep sleep state, and finally as that which transcends these three states as the essence of all levels of consciousness.[12]

Notice in this passage from the Upanishads how the Hindu idea of consciousness (*vijñāna*) differs from the way we usually think of it in the West. For us, someone who is fast asleep is "unconscious"—*not* conscious. It is waking awareness that we think of as "consciousness." In the Upanishads, on the other hand, someone who is fast asleep is conscious, just not of external or internal things. What we would usually call an absence of consciousness is consciousness of nothing. To these Hindus, consciousness is not something that comes and goes; it is always there as the basis of everything. It is integral to the nature of Brahman and the universe.

Not surprisingly, since Brahman is the single aware self of which all aware selves are expressions, Hindus think of Brahman as having certain archetypal qualities possessed by and animating the lives of all aware beings. Nondual Hindus speak of three such qualities: *sat* ("existence," "being"), *chit* ("consciousness," "awareness," "knowledge"), and *ānanda* ("bliss," "happiness").[13] Sat is that whereby we exist and are what we are. Chit is that whereby we are aware of things and their properties and know what we know. Ānanda is that whereby we value one thing and not another and want what we want.[14]

As archetypal qualities of Brahman, they are typically conceived of in an idealized sense related to our purpose in life. Sat is not only responsible for our existence; it is the ideal existence (or identity) we all seek. Chit is not only responsible for our awareness of things; it is the ideal superconsciousness (or knowledge, understanding) we all seek. Ānanda is not only responsible for all value and desire in life; it is the ideal bliss (or most desirable experience or subjective state) we all seek. In the popular imagination, such qualities make Brahman look very much like the concept of God typically found in Western monotheism. Indeed, Brahman is often translated into English as simply "God" in spite of the differences in meaning.

Some readers might criticize Hindus here for anthropomorphizing God. Are they not anchoring their concept of Brahman in our human sense of existence, consciousness, and bliss? Hindus, however, turn intentionally to the human aware self for their understanding of Brahman. The human capacity for intelligence (*buddhi*) allows humans to synthesize many points of view, and at the heart of this capacity is thought to lie Brahman, who sees from all eyes and hears from all ears.[15]

The Everyday World

Brahman is Nondual Hinduism's philosophical first principle and relates to the world of our everyday experience in much the same fashion as the atom archetype I described relates to atom instances. What, then, about the world of our everyday experience, all the particular things and events of our first-person view?

Hindus call this everyday world *samsara*. Samsara is conceived of as an eternal round of birth, life, death, and rebirth, with life a mix of pleasure and pain, good and bad. In other words, it's the world as seen through the mind of an aware self.

In this round of existence, all creatures, not just humans, reincarnate, and one may be reborn into higher or lower life-forms: human on down to insect and perhaps even plant. This concept of "transmigration," or cross-species reincarnation, is critical for understanding Indian philosophy. Its importance lies not so much in whether beings literally reincarnate and become other creatures but rather in what it implies about the continuity of all life. Whatever the differences may be that separate one being from another, these differences are not the core of life, not the underlying reality, not what is ultimately most important. Whatever the most basic essence might be for a human being, that same essence exists in all living beings (recall this chapter's opening quote by Tagore). For the Hindu, you and I are not just human beings; at a more fundamental level, we are kin to all creatures—cattle, bird, fish, even ant. As the nondual school puts it, we are all ultimately Brahman.

One does not have to be a Hindu these days to think of all living beings as related to each other. We are all in certain respects expressions of our DNA, and in terms of shared biological processes, everything that is alive (even plants) shares common properties—a genetic "soul," if you will. For Hindus, however, these structural similarities are themselves expressions of a deeper, more fundamental kinship arising from the nature of aware selfhood.

Placing the aware self center stage implies other things about our everyday world as well. The fundamental laws of nature that govern our lives do not come primarily from physics and biology but rather from the aware self that each of us ultimately is. In other words, we as aware selves are responsible for our own life experiences.

We are responsible for our life experiences in at least two ways.[16] First, we are responsible for who or what we identify ourselves with in life. All of us typically identify ourselves as individual persons having physical bodies, names, social relationships, histories, and anticipated futures. But do we identify ourselves with more than this one particular person we see every day in the mirror? To what extent do we identify with the Ātman? To identify with the Ātman even a little is to identify with more than just the one body-bound self and expand our horizons in ways that offer greater opportunities for fulfillment in life. In fact, most of us do experience ourselves as more than merely our own points of view. We feel pleasure when others feel pleasure; sadness when others are unhappy. To understand how events will unfold, we anticipate as best we can what other individuals and groups will do in various situations.

Second, our life experiences depend on how well our behavior toward others reflects our underlying kinship with others—how well, in other words, we fulfill our ethical duties (*dharma*) and act with moral virtue. This cause-and-effect relationship between our virtue and our fortunes in life is called *karma*. As we sow, so shall we reap. Because we are all fundamentally the same self, our actions toward others redound on ourselves. The choices we make are etched in our own psyches such that our behavior toward others in the past affects our circumstances now, while our behavior now affects our circumstances in the future.

In time, virtue brings more than just worldly rewards. The ultimate goal of life is liberation (*moksha*) from samsara and the eternal round of joy and suffering, birth and death. In other words, the reincarnating soul of a human being is thought to already have eternal life, but eternal life does not mean eternal bliss. To achieve perpetual bliss, we need to escape the cycle of life, death, and rebirth. Moksha is the promise that our moral efforts will eventually pay off and free us from our entrapment in this cycle. It is said that in every generation many people achieve moksha and that it can occur while we are still alive.

Enlightenment

How, then, does one achieve moksha, or liberation? According to the Upanishads, one achieves moksha through a deep and abiding realization that one's true nature is not one's particular mortal being but rather the Ātman that is Brahman. We are each already God; the task is to realize our identity with the divine in our own day-to-day experience and become who we are to begin with.[17]

Moksha is thus a question of *enlightenment*, of overcoming ignorance about one's true nature.[18] On the one hand, moksha is a process of self-discovery—of examining oneself and discovering one's own universal essence. On the other hand, it is a process of habituating oneself to that understanding so that one's identity with God is realized not simply in thought but in practice as well.

At this point, you might have a question for Nondual Hindus. If each of us is already God and God is all knowing, then why don't we know that we are God? In other words, if we are God, why don't we always experience ourselves as all knowing and permanently blissful in the manner of Brahman's own nature? Why, instead, do we experience ourselves as mere mortals? How could God, the all knowing, be ignorant of God's own nature?

Nondual Hindus answer this question with the concept of *maya*, or illusion. Maya, a term shared by other schools, is the ignorance that keeps us from enlightenment. In the nondual school, it is an aspect of Brahman: Brahman's "cloak," or sometimes Brahman's "sport" or "play."[19] In other words, the ignorance that hides from us the true nature of ourselves is part of the way Brahman manifests in the world.

Hindus liken the effect of maya to being asleep and dreaming. When we dream, there is only one dreamer, ourselves; yet in our dreams, we move about in a world of different objects, events, and people. We make friends, have adventures, get into trouble—all in a world that seems real and separate from us, a world not created by us. Only when we wake up do we discover differently. Likewise, Hindus speak of the universe as God's dream. The goal of life is to awaken from the dream. To awaken, we must come to know ourselves as Brahman, the dreamer, instead of characters within the dream.

Maya not only hides our true self from us; it is also the source of all trouble and unhappiness in the world. If we knew ourselves as Brahman, we would see and feel from all points of view, not just our own. Virtuous behavior would be automatic because we would recognize these others as ourselves in different bodies. Because of maya, however, we identify with just our bodies and the body's single perspective. Caring for the community and each other falls apart into egoism and self-interest and then into lying, stealing, and killing. To the extent we identify with our bodies, we

FIGURE 5. Vishnu dreaming the universe (c. 500 C.E.). Relief from Dasavatara Temple, Deogarb, India. *Source*: Joseph Campbell, *The Mythic Images* (Princeton: Princeton University Press, 1974), 6. © Archaeological Survey of India, courtesy of Mrs. A. K. Coomaraswamy.

also face death. In order for us to believe that we live on after our bodies die, we must first identify ourselves with what lives on after our bodies die; we must experience ourselves as something other than just physical or biological beings.

One overcomes the illusion of separateness from Brahman through different spiritual practices called *yogas*. These yogas are not much like the yoga Westerners do to keep in physical shape. There is indeed a component of physical exercises, but much more important are exercises of the mind and the following of moral precepts. These yogas are not philosophical schools. Rather, they are practices that, for the most part, might be applied regardless of one's beliefs about ultimate reality.

These yogas are traditionally divided into three categories (with a fourth, *Rāja Yoga*, often added[20]). The categories are distinguished by the values that motivate us and define our character. Insofar as one values action and productive activity, there is *karma yoga*, the yoga of work and service and of living in accord with one's social duties. Insofar as feelings and affections guide one's life choices, there is *bhakti yoga*, the yoga of love and devotion. Insofar as one is intellectually inclined, there is *jñāna* (jeh-NAH-nuh) *yoga*, the yoga of knowledge (which I'll discuss further in the final chapter).

None of these yogas are exclusive of each other, although we usually lean more toward one than another. Mother Teresa in her care for the poor of Calcutta speaks of devotional practices being central to the service she performs, which from a Hindu perspective implies a combination of bhakti yoga and karma yoga.[21] All these yogas aim at moksha, the uniting of ourselves with God, who, in the tradition of Nondual Hinduism, is the one self, the Ātman.

The Five Cloaks of Brahman

In Nondual Hinduism, yogas can be thought of as spiritual practices that intend to dispel the maya that hides from us our true self, the Ātman. One way to conceive of this maya is as five cloaks, or "sheaths" (Sanskrit: *kośa*), covering Brahman in successive layers.[22] These cloaks provide another of Hinduism's illustrations of how Brahman relates to us as individual beings and the avenue by which self-inquiry brings the realization that we are fundamentally Brahman.

To understand these cloaks, first consider what we directly perceive of another person in our here-and-now sense experience—how sense experience by itself doesn't show us the inner lives or minds of others but rather just physical sensations (sights, sounds, smells, etc.) and outward bodily characteristics and behaviors. When we regard each other like this, we are just material objects, physical bodies. Hindus call this outermost covering the "food-illusion sheath" (*annamayakośa*). Insofar as we think of ourselves and other living beings in terms of our physical bodies, then we are to each other little more than meat and bone, a meal, a material resource.

Second, we can think of ourselves as living beings characterized by nonphysical properties that make us more than simply hunks of matter. We are biological beings, not merely physical ones. This next covering is called the "life-illusion sheath" (*prāna-mayakośa*); it's ourselves as our aliveness or life-force (*prāna*, vital spirit or breath).

Third, we can think of ourselves as more than merely alive. A mass of cells is alive, but we are each a system of cells in which all parts work together as a whole. Each of us is a single organism with its own perceptual and affective perspective—its own "mind," so to speak. This is the "mind-illusion sheath" (*manomayakośa*), or ourselves as an individual way of perceiving and interacting.

Fourth, we can think of ourselves as more than merely our own particular points of view as individuals. Our faculties of intelligence enable us to relate our own here-and-now experiences to the experiences of others and understand the universals and the logic of our world as they apply to all. We can also discriminate good from bad not just in terms of personal interests but in terms of the interests of others and the collective good. This is the "intelligence-illusion sheath" (*vijñānamayakośa*), or ourselves as this capacity to understand the world from a collective perspective that transcends our individual, here-and-now points of view.

Fifth, we can think of ourselves as more than merely the collective view of our intellect. We are actors in the universe propelled toward bliss or happiness or serenity or contentment or whatever we value most. This is the "bliss-illusion sheath" (*ānandamayakośa*), or ourselves as expressing the underlying intention or disposition animating all our individual and collective activities, which includes the activities of all the other sheaths.

In the end, Hinduism conceives of all these sheaths as veils hiding our true Self, the Ātman, which might be thought of as both in and beyond these coverings.

In sum, these cloaks trace a progression in our understanding of ourselves from physical beings to living beings, to beings with their own perceptual and affective perspectives, then to beings marked by an understanding of other points of view, and then beyond even this to that which creates the value driving all choices and actions in our world. In and beyond all these layers of illusion is our true nature: Brahman, the Ātman.

Note that while we today might think of all these layers as ultimately dependent on matter and physical properties, for Hindus, the most important dependent relationship insofar as human nature is concerned is the inverse of this. What is primary is Brahman, on which depends intention (or value), which manifests as the collective logic (or intelligence) that all things follow, which is instantiated in individual points of view (minds), which are animated as life and lastly given form as physical bodies.

Also of note—especially for the intellectually inclined—are the relative positions of the two inmost sheaths. For Hindus, the inmost veil hiding Brahman is not that of our intellect but rather that of our *values*. These dispositions and interests express themselves not only in our choices and actions but also in the rationale our human minds find in or impose on the world around us. They also, as I mentioned in the previous section, inform our spiritual practices—the mix of karma, bhakti, and jñāna yogas best suited to each of us for achieving moksha.

The Different Philosophical Schools

Hinduism was far from unanimous in its views about ultimate reality and divided into a number of different schools of thought that continue today. The idea of Brahman as the sole principle of nature is developed by the "nondual" and "qualified nondual" schools (*Advaita-Vedānta* and *Viśiṣṭādvaita-Vedānta*). Two other schools, *Sāṃkhya* and *Yoga*, also speak of the conscious self as a first principle, but for them, there is no underlying supreme being, or Brahman. In these two schools, the conscious self is inherently many; it is what makes each of us a separate first-person view, an aware individual. We are each in essence a single, unblemished occasion or presence of awareness (*puruṣa*) without ego or desire. What animates us and causes action is our relationship to matter (*prakṛti*), which is also a first principle in these schools. All things in the universe arise out of the interactions between innumerable conscious beings and matter, with matter encompassing everything else in the universe apart from conscious selves, including the rules that govern the universe's orderly behavior. The goal of human life is to quiet the effects of matter on ourselves and discover ourselves to be the pristine consciousnesses that we are to begin with.

There are other schools of thought as well. But, as in the West, the general population is less concerned with such philosophical differences than with spiritual practices and the benefits that might accrue from them. In practice, most people prefer to think of Brahman as distinct from themselves and the world in a fashion very similar to the God of Western monotheism (the view developed by "Dual" Hinduism, the *Dvaita-Vedānta* school). In popular worship and ritual, philosophical first principles recede behind the less-abstract faces of deities like Shiva, Vishnu, and Durgā (the divine mother; consort of Shiva) or mythological God-realized individuals like Rāma and Krishna, who play roles very similar to that of Jesus in Christianity. Everyday goals of worldly success and progeny, satisfying desires, and living a worthy, virtuous life are at least as if not more important than moksha. Then, too, the literal interpretation of scripture tends to be more important than personal revelation and enlightenment.

Summary with Strengths and Weaknesses of This Strategy

Hinduism's nondual school has taught one of the most enduring and successful third-person accounts. Many Indian philosophical schools assume that reality is not something to be known apart from the awareness that knows it. Building on this idea, the nondual school traces all that we experience back to a single principal, the aware self, or Brahman, from which arise all aware selves and indeed everything in the universe, including matter. The relationships among aware selves and between each aware self and Brahman are thus seen as the principal dynamic of our everyday life, a dynamic expressed in such concepts as karma, the moral law of cause and effect that governs our life circumstances.

Another aspect of this dynamic is the impetus of all aware beings toward moksha, or liberation from samsara, the everyday round of joy and suffering. In Nondual Hinduism, moksha is won through the realization of one's identity with Brahman, which in everyday life is concealed from us by maya, or illusion. This realization is accomplished through yoga practices, which are traditionally divided according the values that motivate us and define our character. Three principal yogas are karma, the yoga of work and service; bhakti, the yoga of love and devotion; and jñāna, the yoga of knowledge.

Maya can be conceived of as five cloaks that hide from us our true Self, the Ātman. The outside covering is our physical bodies; within that is the life-force, what makes us alive and not inert; within that is our individuated awareness, our minds; within that is our intellect; and within that is ānanda, our inner sense of and disposition toward what gives us "joy" (the underlying affective aspect of ourselves that motivates choices and actions).

Hinduism also embraces many philosophical schools, of which the nondual school is but one.

This idea that we are all expressions of a single archetype that is more truly us than our individual, mortal selves has had enormous appeal in both the East and the West. Most mystical traditions associated with Judaism, Christianity, and Islam share a similar idea (which I will touch on later).

In spite of the historical appeal of this approach, it seems difficult these days to understand our universe in terms of such a principle. The vast bulk of our universe seems to be made of inert matter, not aware selves. Granted, the Upanishads have a different idea of awareness then we do—deep sleep is consciousness of nothing. But it is difficult to understand what they might mean by such an idea. The human waking awareness we call "consciousness" seems tied to the complexity and sophistication of the human brain; if you don't have a brain or the brain stops working, you don't have consciousness. How, then, can consciousness be the origin of all things when it seems itself to arise from what is not conscious?

Nevertheless, in weighing such arguments, it is important to remember that the universe of interest to Hindus here is specifically the universe that we humans are aware of, and from this perspective, nothing is more fundamental than the aware self. Also, as I've discussed, the mind's relationship to matter has been and remains a troublesome problem for Western philosophy, one that many philosophers historically have chosen to resolve by finding their first principles in the mind or awareness.

But even if we were persuaded that the mind or awareness might provide a first principle, we are still left with the problem of maya, or illusion, as an aspect of this first principle. In the context of Nondual Hinduism, how specifically does maya hide from us our identity with Brahman? How can Brahman, the all knowing, hide itself from itself? In fact, the nondual school says that maya is inexplicable.[23] (I will come back to this last difficulty in later chapters. It has a striking symmetry with difficulties found in the other philosophical approaches I will look at.)

Such challenges to Hinduism's nondual school of philosophy point to a problem with third-person accounts in general. As compelling as some of these accounts are, there always seems to be something incomplete or contradictory about them, or something we are asked to accept on faith. East and West, we have debated for two and a half *millennia* over which universals of mind or matter or God best illumine the nature of the third-person view, and the issue still remains unresolved.

To me, this interminable debate is itself evidence that third-person accounts— at least by themselves—are inadequate for fully understanding reality. As Nondual Hindus themselves claim, the ultimate reality of Brahman is beyond naming; it cannot be captured in universals.[24] It therefore seems to me that the biggest challenge to any third-person account is not which universal or universals best describe the nature of reality but whether *any* such universals are, in the end, sufficient by themselves to describe it.

This challenge to third-person accounts brings me to my next subject.

CHAPTER FIVE

Awareness and Its Objects

First-Person Accounts of Reality

Critics of accounts of reality anchored in universal properties of the cosmos—critics of what I'm calling third-person accounts—have been with us since the beginning of philosophy. A Greek by the name of Cratylus (sixth century B.C.E.) purportedly thought the universe so particular and changeable in nature that it could not be described at all. He eventually stopped talking and, according to Aristotle, communicated only by moving his finger.[1]

While some like Cratylus based their criticisms of third-person accounts on the ever-changing and transitory qualities of nature, others had different problems with these accounts. Pyrrho the Skeptic (c. 365/60–c. 270 B.C.E.) argued that our faculties of knowledge, specifically perception and logic, could not be trusted to show us universal truths. A century earlier, Gorgias the Sophist (c. 483–c. 376 B.C.E.) argued that it was as easy to prove that "being" (the "is"-ness of everything that *is*) had no existence at all as it was to prove (per Parmenides) that being was a first principle of nature.[2]

What most of us want in an account of reality, however, is more than merely a criticism of others' efforts. If we cannot satisfactorily understand reality in terms of universal first principles, then where else might we turn to understand it? One place many have turned is to the first-person view—an individual's own awareness of a world.

In a sense, this is where Hinduism turned. You might have noticed that Nondual Hinduism's third-person account begins with a first-person view, an aware self—the ātman that is Brahman. What makes Nondual Hinduism a third-person account instead of a first-person one, however, is that it treats the aware self as a first principle, an archetype. Brahman is that in which first person and third person are one, but as a universal.

The alternative approach that I'm interested in now is to leave the first-person view as we actually experience it in our everyday lives and treat it as particular in nature rather than universal. In our daily interactions in the world, we do not typically see ourselves as instantiations of a first principle like Brahman—as the supreme being in human form. From my first-person perspective, I experience myself as a particular person. I am me, and you are you; I see what I see from my point of view, and you see what you see from yours.

In a first-person account, this stress on particulars over universals extends to the nature of reality itself. To use the atom instance versus atom archetype illustration, such a strategy highlights the atom instance side of the ledger. The reality we seek to understand is *fundamentally* many, not one; it is *fundamentally* space and time bound, corporeal, finite, and causally interactive. While universals exist, they are derivative, not primary. They are created by us; they don't preexist in nature.

What, then, might a first-person account of reality look like?

There are many varieties of first-person accounts; however, most of our more recent Western versions trace historically to the phenomenology developed by Edmund Husserl in the beginning of the twentieth century. Husserl's phenomenological approach begins with an individual's awareness of a world and then proceeds to a description of the basic properties and structures that characterize this awareness—not just for oneself but for others as well. In other words, such accounts describe oneself being aware insofar as we observe it to be any individual being aware.[3]

The issue raised by phenomenology, therefore, is the structure of first-person awareness from the vantage point of the aware individual. We all know roughly how awareness works from a third-person, physiological perspective. In the case of visual perception, for example, photons strike the retina; nerve impulses pass to the brain; the visual cortex interprets the data. But this does not describe our own first-person visual awareness. We don't see photons or nerve impulses; rather, we see the book we are reading, people, cars, trees, stars. We see the results of the perceptual process—results that don't look at all like the inside of a brain.

I will shortly illustrate such an account. First, however, I want to fill in some historical details prior to Husserl.

The Cartesian Roots of Western First-Person Philosophy

After the decline of Greek and Roman philosophical inquiry, perhaps a thousand years passed before anyone in the West again claimed that universals were invented by the human mind. There were mystics and Gnostics who associated universals with a Brahman-like archetype of human intelligence, but this concept of intelligence was grounded in the third-person view, not the first-person view. This was Human Intelligence with the capital *H*, not the small *h*.

The long-undisputed reign of mind-independent universals was challenged at the end of the eleventh century by Roscelin de Compiègne (1050–1125) and his student, Peter Abelard (1079–1142).[4] Specifically at issue was Aristotle's idea of universals as natural features of the universe. Roscelin argued that such universals were merely words we speak, or labels we apply to the world (the view today called "nominalism"). Abelard, like Roscelin, believed that universals were constructions of the mind, but his position was less extreme in that the world did have common features that justified being described with universals. Abelard's more careful analysis of how the mind constructed universals through a process of abstraction greatly influenced subsequent views.

After Abelard came a variety of positions. Abelard's student, John of Salisbury (c. 1120–1180), for example, followed Abelard's more moderate course, while St. Bonaventure (c. 1221–1274), St. Thomas Aquinas (1225–1274), and John Duns Scotus (c. 1266–1308) held the traditional Platonic and Aristotelian view of universals as mind independent (i.e., as "real").

William of Ockham (c. 1285–1349) ended the controversy over the reality of universals for a while. With more sophistication than Roscelin, he argued that all universals came entirely from the mind but, instead of being mere words, were generalizations made from experience (the view today called "conceptualism"). Properties like "being human" resided not in nature or in words but in the concepts people create to describe what they observe. Ockham supported his position by pointing out (as Plato did) that sense experience reveals only particulars, not universals themselves. For Ockham—unlike Plato—the human mind in interaction with the world creates universals in order to organize the particulars of sense experience.

With Ockham, the idea that universals might be mind independent fell into disrepute, as did people like Duns Scotus, who had supported this view (the word *dunce* comes from "Dunsman," an advocate of Duns Scotus's philosophy). In keeping with the new humanism of the Renaissance, the human mind was seen over the following centuries as the source of universals and the means by which the things we observe came to have their distinctive properties.

This newly established and pervasive Western focus on the human mind as the basis for reality as we know it set the stage for René Descartes (1596–1650 C.E.), who completed the philosophical transformation that had started in the late Middle Ages. Ockham had finished moving universals into the mind, but the particulars of the world we saw around us were still outside and real. Descartes moved particulars, too, into the mind. All of reality became something apart from what any person directly perceived, something we might know only through our minds and reason (if at all). Before Descartes, the sun that each of us saw was the real sun, the same sun everyone saw, though how we each might see it and think about it could be very different. After Descartes, the sun one person saw was the appearance of the sun in that person's mind, while the sun others saw was the appearance of the sun in their minds.

Accompanying Descartes's insight was a new way of doing philosophy. A mathematician as well as a philosopher, he invented Cartesian coordinates and was the principal developer of analytical geometry. Like many others before him, Descartes saw in mathematics a realm of certainty not available in the world of sense experience. Descartes wanted to put philosophy on a similar firm foundation—one that, like math, could not be doubted.

In his most famous philosophical work, *Meditations on First Philosophy*, Descartes wrote that such certainty could be found within the very nature of human self-reflection. We can doubt the truth of every philosophy, doubt the reality of every perception, and even doubt the existence of the world and our bodies; after all, we could be dreaming or hallucinating. But what we cannot doubt in the midst of such reflection is that we are having these doubts. We can't doubt the doubt or ourselves, the thinking mind, as the arena of that doubt.

Beginning with this small purchase of certainty, Descartes examines the landscape of what he has discovered. Within this inner realm of certainty, he finds himself to be "a thing that thinks; that is, a mind, or soul, or intellect, or reason. . . . a thing that doubts, understands, affirms, denies, wills, refuses, and which also imagines and senses."[5] As he had concluded in an earlier work, *Cogito ergo sum*—"I think, therefore I am."

But what about things outside this inner realm of certainty, his mind? What can he be certain about in regards to the objects he perceives around himself?

Descartes tells of himself sitting before a fireplace with a piece of fresh honeycomb wax in his hand. As he brings the wax close to the fire, its perceptual properties change almost entirely; it becomes pliable and changes shape, color, odor, and even sound when tapped. The only way he knows that it was and remains a piece of wax is that his mind judges it to be so. But if the wax or anything else that he perceives is just a judgment made by his mind, then how can he be certain that things outside his mind truly are as he perceives them? In fact, he cannot be certain.

In this way, Descartes withdrew philosophical certainty from the third-person view and placed it in the first-person view. For Descartes, philosophy begins with a person's own awareness of mind and world and a person's doubt as to one's ability to directly experience a true-for-all, third-person world external to the mind.

Descartes's philosophy is not, however, what I call a first-person account of reality. Reality for him is still a mind-independent, third-person universe, though one we cannot directly experience. Nevertheless, he paved the way for first-person accounts as we think of them today. He showed us a manner of doing philosophy that begins with our own first-person view.

I've discussed elsewhere some subsequent efforts to make sense out of Descartes's gap between interior mind and certainty and what is exterior to the mind. For this section, I want to skip to Edmund Husserl (1859–1938) and phenomenology.

Husserl explicitly built on Descartes's emphasis on the first-person view, but he removed any judgment as to reality's mind independence. He developed a method of philosophical inquiry wherein one puts aside ("brackets") all questions about mind-independent existence in order to examine strictly the first-person contents of one's own consciousness.[6]

For example, imagine a person sitting in a room and looking around. The things the person sees could be real, mind-independent tables and chairs. Or they could be neurological representations of what really exists. Or, again, they could be imaginary products of a daydream. Or, still again, they might even be hallucinations. Husserl asked us to put aside questions like these that are related to their state of existence and describe just the room as it is present in our minds. He showed that such an analysis reveals certain logical structures seemingly present for everyone, at least everyone we presume has a first-person view like our own.

The next section illustrates such an account. Besides giving an example of a first-person approach, it also provides a post-Descartes context for better understanding the Buddhist philosophy discussed in the next chapter. Indeed, many writers have turned to phenomenological analyses to better explain key concepts in Buddhism.[7]

Not all phenomenological accounts are presented as a narrative like this, and I've tailored this one to my own purposes; my structures of consciousness are not exactly the same as those Husserl emphasizes. I also don't have cats anymore, though I've added one here to help me out.[8]

A Phenomenological Illustration of First-Person Philosophy

It's early morning, and I'm sitting in my study at home with my cat, Merlin, whom I have just pushed off my lap. While Merlin roams around the room, I settle in my chair to examine my first-person awareness of a world.

To begin, I put aside the thought that there is a third-person view independent of my own mind. Putting aside this thought of a mind-independent reality doesn't change anything that I see around myself. Everything in this room that I had been conscious of before remains now: the desk in front of me, my computer, books in bookcases along the walls, the window with a window seat and hanging plant, and Merlin. The only difference is that I now consider them strictly as they are presented in and by my awareness.

My focus turns to a pen laying on my desk. When I look at this pen, what do I observe about the nature of my awareness? What sort of structure does my first-person awareness of a world have when I reflect on this perceptual experience?

Its main structure is "I . . . see . . . this pen." My awareness encompasses not only the act of perceiving but also what I perceive as well as myself as a perceptual perspective. This characteristic structure is little more than a restatement of what the first-person view is: an individual being aware coupled with whatever is present in that awareness.[9]

I turn my attention from the pen to the other things around me. What else might I say about my awareness of these things?

One thing I observe is that no matter where I look, I'm always aware in the here and now. If I shut my eyes, then open them again, it's still the here and now. If I set this alarm clock on my desk for any time at all, whenever it goes off, it's still the here and now. If I imagine myself in some far-off past or future time, the imagining itself still takes place right here and right now.

Whatever it means to me to be here and now, it's my first-person awareness that gives it that meaning. It's from my personal here and now that I look forward to the future, back to my past, and outward to whatever may be present in space before me.[10]

So in this fashion, my first-person account of reality starts to have a structure. Its foundation is an individual, an "I," being aware of various things (whatever is presented), with this awareness always taking place in the here and now. This foundation contrasts sharply with Hinduism, where Brahman is that who sees through *all* eyes and is aware from *all* points of view at the same time.

A knock on my door interrupts my meditation. My wife comes into the room to say good-bye before leaving for work.

After she goes, my reflections turn to other points of view besides my own. Among the things I'm aware of are people like my wife, whom I think of as also aware of various things. What part do they play in my first-person account of reality?

To begin with, I don't directly perceive the minds or the first-person views of others. When I look at others, I perceive behaviors and responses to behaviors—the causal interactions that describe the relationship of one particular thing or event to another. I perceive how objects look when light shines on them; how they sound and smell; and if they speak, what they say. It's on the basis of such interactions that I draw conclusions about and assign properties to the things I perceive around myself. And one of my most important and far-reaching conclusions is that there are others like myself who have first-person views.

This conclusion that there are other first-person views besides my own shows up in the properties of the things I perceive around myself; I don't think of them as things that only I can perceive. My chair is for me not only something that I can sit on; it is also something that others can sit on. My books are not only things I can read and think about; they are also things that others can read and think about, things I can also talk about and discuss with other people. Likewise with the clock, desk, computer, pictures, and other objects in this room. I experience all these things as having properties that not only I'm aware of but others seem to be aware of too—properties that are *public* and belong to a world shared with other aware beings.

Of course, these public properties are not the only properties I associate with these things. There is, for example, a seashell sitting on my bookcase that I brought back from California many years ago that carries personal memories of that period in my life. But I don't confuse such private properties as these with the public ones that others might perceive—or at least I try not to confuse the two.

My awareness does something else, then, as it identifies the things in this room: it distinguishes as far as possible those properties that are public from those that are not. In other words, it makes a distinction that is very much the same third-person versus first-person distinction that I spoke of earlier, except that here "third-person view" doesn't mean "mind independent." In this case, the third-person view is anchored in first-person awareness insofar as I am aware of other individuals with first-person views who perceive much the same things that I do.

To summarize the structure so far of my first-person view, its basis is me being aware of various things, with this awareness always taking place in the here and now. In this awareness of various things, my first-person view distinguishes private from public properties. It distinguishes properties that apply to me alone from properties that presume the existence of other first-person views that are aware of much of what I am aware of. It also distinguishes from other things those individuals (such as my wife) who appear to have first-person views that are similar in many ways to my own.

Merlin jumps back on my lap.

I scratch Merlin behind an ear. The cat tilts his head against my fingers as I continue scratching.

Perhaps Merlin sees some of what I see in his own cat sort of way; certainly the window seat is there for him when he wants to take a nap. But how much of what I see in this room do I really think Merlin also sees? To him, my books and pictures seem to be little more than just curious hunks of matter. My pens and pencils are more toys to play with than things to write with, and I very much doubt that he sees my books as things to read or my clock as a way to tell time or my pictures as works of art.

Reflecting on Merlin's point of view as well as the points of view of other living creatures reveals another feature of my awareness. My first-person awareness discriminates *differences* in what different individuals might be aware of. It presents this desk, computer, books, and so on as things that Merlin and I *don't* perceive in quite the same way. Nor do I see eye to eye with the spiders and other bugs I find in here. Moths certainly have a different view of my sweaters than I do.

So my awareness of public properties is more complex than I characterized it a moment ago. My awareness doesn't simply distinguish public from private; it sorts out as far as possible who is aware of what. It sorts out whose awareness is enough like mine to perceive what I perceive, as well as who might perceive very different things.[11]

I look down at Merlin, who has begun to purr.

No doubt with Merlin in my life, my lap is now more than just the top of my legs when I'm seated. And my desk has picked up the property of being a place where a cat can play with pens and pencils. Such properties as these are also now part of my public world, as is Merlin himself and his different way of being aware.

But properties that depend on the various similarities and differences in the ways living beings are aware are not the only properties I regard as public. In the case of my desk, when I put aside its aesthetic appeal (such as it is) or the various ways it might be useful to a human or a cat, there still seem to remain other properties, ones I associate not with the awareness of living beings but rather with matter: the desk's shape, its spatial location relative to other things, the properties of the wood out of which it is made. Physical characteristics such as these seem to be here in some fashion for all creatures. It is on top of or in relation to such qualities that I and others seem to add our own properties—ones particular to our different first-person views.

Thus the pen I see before me has physical properties, such as its material composition and spatial location, that appear to be here in some way for both Merlin and myself (even when we're not aware of them). Additionally, it is for me a writing instrument, while for Merlin, apparently, it is an occasional plaything. Thus my sweaters have the physical properties of wool that seem to be here for both moths and myself, but when I look at wool, I see something warm, while moths apparently perceive a meal.

My phenomenological analysis has now led me to properties of things in this room that do not appear to derive from or depend on the awareness of any living beings; they appear instead to come from the physical materials out of which these things are made.[12] We have, in other words, come by a different route to the distinction between mind and matter. However, in a first-person analysis such as this, I don't consider these seemingly true-for-all material properties to be mind independent (I've put aside considerations like this). Instead, they describe what appears to me to be true for all first-person views (which coincides with, it seems to me, the instrumentalist account of our physical universe).

Regardless of these different kinds of properties, however, the most important properties by far are those that come with the way of being aware I share with other humans. It's mainly, though not entirely, by these properties that I distinguish the objects around me as a cat or a desk or a pen and decide how they might fit into the life of a person such as myself.

In sum, my first-person account of reality begins with me being aware of various things, with this awareness always taking place in the here and now. In the process of distinguishing among particular things in the world, my awareness sorts out who is aware of what. It distinguishes what might itself be aware and have a first-person view as well as what properties might be public and associated with these various first-person views. It also sorts out what properties are present but not associated with any apparent first-person view, properties that seem to hold for everyone and are typically associated with the materials out of which things are made.

One more feature needs to be added to this account of reality. My first-person awareness doesn't just passively experience various things; it responds and interacts. The fundamental dynamic that I observe in the public world is cause-and-effect interaction, the kind of causality that relates one particular to another. And in my awareness of a world, I participate in this interaction, as do cats, other people, and the material objects I see around me. We are not only aware of others and their behavior; we produce behavior that others are aware of. And out of these interactions come what happens next.[13]

Awareness and Conscious Awareness

The concepts of awareness and consciousness play central roles in phenomenology and translate to concepts that play central roles in Asian philosophy (recall Nondual Hinduism's first principle of Brahman as the aware self). But before discussing Buddhism, I want to draw your attention to some of the ways we in the West typically use these two words in everyday speech.

First, *awareness* is the more general of the two words. When we use *awareness* and *consciousness* together with one modifying the other, we say "conscious awareness" and "unconscious awareness," not "aware consciousness" or "unaware consciousness." It is consciousness that modifies awareness, not the other way around.

Second, in keeping with its more general usage, we tend to think of awareness as coming in many varieties, of which conscious awareness is but one. We generally speak of sleep, for example, as a time when we are not conscious yet are still aware enough to wake up if we hear a loud noise. Even when we are awake, we are unconsciously aware of many things. If someone unexpectedly tosses us an object, we will respond to catch or avoid it well before we're conscious of what we're doing. We see this unconscious awareness also in the "cocktail party effect." If you're at a noisy cocktail party and engaged in a conversation, you will typically filter out everything except that conversation. You won't be conscious of other conversations or background noises. Nevertheless, some part of your awareness still monitors them. Should someone mention your name, for example, that conversation will suddenly become conscious to you.[14]

Awareness in this general sense is also something we tend to think of as not confined to human beings. Animals are, as we say, "aware" of their circumstances. Some are aware of subtle differences in smells; others are aware of sounds humans can't hear. Consciousness, on the other hand, we usually think of as our own particular human variety of awareness. While an ant might be aware of different smells sufficiently to use them in a rudimentary language, we do not usually think of ants as possessing consciousness. The debate over whether chimpanzees or dolphins or elephants have consciousness is a debate in part over the extent to which their forms of awareness resemble our own.[15]

If consciousness tends to refer specifically to a certain variety of awareness that humans have, what, then, might distinguish consciousness from other sorts of awareness?

One of the most popular and long-standing theories on this matter (and there are many such theories) holds that consciousness involves a "higher-order" monitoring of lower-order internal mental states and processes. Thus, for example, I may always be capable of hearing my refrigerator motor when it is running and I am in the kitchen, but most of the time I'm not conscious of the sound. I become conscious of it only when I notice it or am aware of hearing it. In other words, even though I always have some awareness of the sound (if it suddenly became irregular, I would likely notice), that sort of awareness is not by itself conscious. It is only when some higher-level awareness is brought to bear that it becomes conscious.

There are many versions of higher-order ("HO") theories, but they all have in common this double-tier structure. It is in this sense that one speaks of human consciousness as the capacity to be self-aware—to be able, for example, to reflect on our own process of reflection, to know oneself as an aware being.[16]

In sum, we tend to speak of "awareness" as coming in many varieties, while "conscious awareness" usually refers to the kind of awareness we humans have when we are awake, and then only to a certain kind of attentive or focused awareness (i.e., not the "unconscious" awareness we might have of, say, a refrigerator running in the background). According to one long-standing theory, consciousness is distinguished by a double-tier structure where there is awareness of one's own awareness.

This distinction between conscious and unconscious awareness is especially important in my context here.[17] As indicated by our many theories on the subject (e.g., the "HO" theories), it has been of considerable empirical and philosophical interest. It also plays a central role in my later discussions of both awareness and personal identity—what it means to be an aware self.

These ways we in the West talk about awareness and consciousness are also important to keep in mind when discussing Hindu and Buddhist philosophies. Recall the Sanskrit word for what is usually translated as "conscious awareness"—*vijñāna*—and how its meaning in Hinduism differs from what we generally mean by conscious awareness in the West. Nondual Hindus do not think of vijñāna as something that comes and goes depending on whether we're awake or asleep. Vijñāna underlies everything; it is in the contemplation of our own aware selves that we discover the Ātman, the one aware Self that underlies all that exists.

Buddhism, on the other hand, takes another tack. As I'll discuss shortly, awareness plays a fundamental role here as well; however, *conscious* awareness, or vijñāna, is viewed more in the way we might view consciousness in the West. It is considered to be derivative—a component of first-person experience that arises from and is shaped by causal conditions.

As you consider these differences among various concepts of awareness, here is one more observation. Since the Enlightenment, we tend to think of awareness as something interior to our individual minds, something like joy or sorrow or other subjective experiences that cannot be perceived from the outside by someone else. The Cartesian gap between interior mind and exterior matter gives us good reason to think of awareness like this; what in fact do my senses directly show me of another's awareness except perhaps physical indications suggesting its existence? Nevertheless, much of our first-person experience presumes a relationship to the awareness of others that is much more intimate than this. We experience this intimacy, of course, in our affections for others, but recall also my phenomenological analysis of awareness. We experience the world around us as having properties that depend not only on the awareness of others but also on our having sufficient access to the awareness of others to judge, to some extent, who might be aware of what.

Summary

As with third-person accounts of reality, the Enlightenment transformed the character of Western first-person accounts. Before Descartes, first-person accounts argued against only the mind-independent reality of universals. No one doubted that the particulars of our perceptual experience showed us a true-for-all world outside our minds. Since Descartes, however, it has become difficult to understand how we can be directly aware of anything outside our minds, including particulars, and first-person accounts of reality have changed accordingly. Most accounts these days derive from Husserl's phenomenology and typically start with an individual's own awareness of a world, but without any judgment as to the world's mind independence.

My illustration of a phenomenological analysis noted certain logical structures seemingly shared by any awareness insofar as it resembles my own. It always takes place in the here and now. In distinguishing among the things it perceives, it sorts out as far as possible what might have a first-person view, what properties might be public (i.e., the same for a group of first-person views), who might perceive which of these public properties, and what properties might be attributed to matter instead of living beings. Awareness is also interactive; it participates in the moment-by-moment causal dynamic that characterizes all particular things.

We typically speak of awareness as coming in many varieties, human and non-human. Conscious awareness, on the other hand, usually refers to just one variety of awareness—namely, human waking awareness, and then only to a certain attentive or focused kind of waking awareness (e.g., we can be aware of certain background noises without being consciously aware of them). Conscious awareness is distinguished according to one long-standing theory by its capacity to apply a "higher-order" awareness or focus to certain of its own awareness activities.

Since the Enlightenment, we tend to think of awareness as interior to our minds even though we live our day-to-day lives presuming to know the awareness of others sufficiently to judge, to some extent, who might be aware of what.

CHAPTER SIX

Buddhism and the First-Person View

Our true home is the present moment. To live in the present moment is a miracle. The miracle is not to walk on water. The miracle is to walk on the green Earth in the present moment, to appreciate the peace and beauty that are available now. Peace is all around us—in the world and in nature—and within us—in our bodies and our spirits. Once we learn to touch this peace, we will be healed and transformed. It is not a matter of faith; it is a matter of practice. We need only to find ways to bring our body and mind back to the present moment so we can touch what is refreshing, healing, and wondrous.

From Touching Peace: Practicing the Art of Mindful Living (1992) by Thich
Nhat Hanh (b. 1926), Vietnamese Zen Buddhist, founder of the Order
of Interbeing, peace activist (Nobel Peace Prize nominee), author[1]

Why Buddhism?

I have turned to Buddhism to illustrate a first-person account of reality for a couple of reasons. First, it is one of the most widespread religions today. It plays a central role in the religious lives of people living in China, Japan, Korea, and much of Southeast Asia, and it has found a large, growing audience in the West.

Second, although Buddhism was founded in fifth- or fourth-century B.C.E. India within a culture dominated by early Hinduism, its teachings depart sharply from those of the Vedas and Upanishads. Buddhism's founder, Siddhartha Gautama or, as he is commonly called, the Buddha (Sanskrit/Pāli for "awakened one"), did not teach that reality had an underlying universal nature such as Brahman. On the contrary, he taught that all things including ourselves were impermanent; nothing had an enduring self or soul. What existed instead were particulars and their interactions—what is present to us in our first-person views.

Like Hinduism, Buddhism has a great variety of different beliefs. The following account focuses on Siddhartha's main teachings as found in its oldest scriptures, the Pāli Canon.[2] However, I will also touch on Buddhism's historical development, as certain later concepts are helpful for understanding issues associated with first-person philosophical accounts.

The Buddha and the Four Noble Truths

Siddhartha lived in India during roughly the fifth century B.C.E.—a time of social and political instability. Small feudal kingdoms had started to give way to larger ones; within perhaps a century, the district where he lived would be swallowed up within the Magadha Empire.

It was also a time of religious and philosophical instability. The Hindu priests, the Brahmins, controlled Indian spiritual life, but with many dissenters. Wandering ascetics lived in the forests away from society and religious orthodoxy. Usually traveling in groups under the tutelages of religious teachers, they debated the ideas found in the Vedas and Upanishads and experimented with new spiritual practices. Within Hinduism, the number of Upanishad texts grew. Outside Hinduism, other religious alternatives were explored. Not only Buddhism but another major world religion, Jainism, developed in India at this time.

Most of these teachers accepted much from Hinduism, as would Siddhartha. Saṃsāra, for example, was humanity's obvious mortal predicament. Clearly, people were born into an imperfect world, experienced joy and suffering, and then died. Nor did these teachers usually question reincarnation or karma as aspects of saṃsāra. One life followed another, with the events of each life caused by prior choices and actions. Instead, these itinerant teachers questioned other things, such as the underlying nature and origin of this reality of saṃsāra and karma, the cause of suffering, the goal of life, and the spiritual practices that could achieve these goals.[3]

Siddhartha would himself become one of these wandering ascetics; however, he began his life under very different circumstances.

The traditional story tells of him being born into a royal family in Northeast India, where he led a very comfortable life, married, and had a child. Although his parents tried to shield him from every physical and psychological distress, at about age twenty-nine, this insular bubble around him broke. In what is called the story of the "four passing sights," we are told that Siddhartha saw for the first time a person who was old, then a person who was diseased, then one who was dead. Last, he saw a wandering ascetic who seemed entirely at peace with the world.

Troubled and provoked by what he saw, Siddhartha left his home, wife, child, and parents to become, himself, a wandering ascetic seeking truth. He traveled for six years, learning and practicing various spiritual disciplines and even trying extreme physical privation. Eventually, close to starvation and still unenlightened, he concluded that the way to salvation was not through weakening the mind and body. Taking some milk rice offered by a cowherd woman, he ate, bathed, and then sat under a tree, resolving not to move until he achieved enlightenment.

After difficult meditative struggles, he attained the awakening he sought—one that would eventually influence almost half the world.

Significantly for the religion that would grow from his experience, it traces its origin not to a body of sacred texts as does Hinduism but rather to hard-won personal insight. At the heart of Buddhism is the Buddha himself, teaching by example that the highest truth is not to be found in words or scripture but in one's own direct experience.[4]

After Siddhartha's enlightenment, he sought out the five ascetics who until recently had been his traveling companions. In his first sermon, given in the Deer Park near Benares, he taught them the "middle way" between extremes that he had learned from his own experience. One does not win enlightenment by indulging one's desire, such as Siddhartha himself did in his youth, but neither does one win enlightenment by ascetically denying basic needs as he did when he was traveling with his five friends.

He then taught them the Four Noble Truths, which remain today the core of the Buddha's teachings.

The first truth is that life is suffering, or *dukkha*. Our lives inevitably come with birth, aging, disease, death, sorrow, grief, and despair. One loses or doesn't get what one wants in life; one gets what one doesn't want. What pleasures and joys we do experience are temporary and, like every other experience, pass away. Suffering in the Buddha's sense is more than just how we feel when bad things happen; it is built into life as part of sentient existence's impermanent nature.

The second truth is that suffering is caused by habituated desire, or *taṇhā* (literally "thirst"). We are attached to pleasurable things and circumstances. These attachments (or "cravings" or "clingings," as they are sometimes called) drive the engine responsible for dukkha. Taṇhā presses us to achieve as much pleasure, joy, and happiness as we can and avoid as much pain and sorrow as we can, and in so doing, it turns the pleasure-pain treadmill of our saṃsāric day-to-day lives.

The third truth is the truth of *nirvana*, the cessation of suffering that comes with ending taṇhā (attachment). Because suffering is caused by taṇhā, ending taṇhā ends suffering. More than that, according to the Buddha, in ending taṇhā, one achieves bliss, even in one's current life.

The fourth truth is the truth of the means to end taṇhā, the eightfold path of spiritual practices that leads to nirvana.

At the center of the Buddha's teachings, therefore, is not an archetypal aware self like Brahman. Instead, there is a practical strategy for overcoming mortal life's built-in difficulties without recourse to concepts like God, Ātman, or Brahman. This pragmatism extends to philosophy. He refused to answer certain philosophical questions, such as "Is the universe eternal?" or "Is the soul the same as the body?" or "Does one live on after death?" According to him, we overcome suffering not by dwelling on questions like these but rather by focusing on the life we actually experience—on what, in other words, our first-person view shows us.[5]

The Nature of Reality

While the Four Noble Truths may give the heart of the Buddha's message, more immediately relevant to my discussion here is the way he characterizes existence. He describes it using three traits or "marks" that are indicative of first-person philosophical accounts.

The first of the three traits is *impermanence* (*anicca*)—there is no constancy in what we experience; all that exists is changing. The second trait is *no-self* (*anatta*; literally "no ātman")—humans have no enduring self or soul. Like the rest of existence, humans are changing too; there is no inner permanent essence. Nor does the cosmos as a whole have a fundamental nature; there is no Brahman or supreme being like the God of monotheism. Spiritual beings—gods and demons—exist, but they, too, are creatures of samsara and change. Third is *dukkha*, which refers to the subjective component of first-person experience and encompasses the entire gamut of feelings, both pleasurable and painful, that make up the "win some, lose some" cycle of life. Life changes and is fragile; even pleasure is dukkha because it cannot last.[6]

Thus instead of Brahman or ātman, the Buddha stresses the impermanence of phenomena and the absence of any ultimate enduring nature or enduring joy in anything attached to this passing drama of life. To show us reality, the Buddha points not to underlying universal properties of the cosmos as a whole but rather to what is actually *present* to us in direct experience. And all that we directly experience are particulars and their interactions.

Notice what the Buddha says here about the "self" or "I" of the first-person view. In my phenomenological account in the previous chapter, I began with an "I" being aware of various things. The Buddha points out that this "I" of our first-person view doesn't refer to an enduring being—someone who was born on a certain day and has had innumerable experiences over the years. Rather, it refers to our here-and-now occasion of awareness—the self of our *present-time as-lived* experience.

If, however, we are only what we are right now and have no enduring nature, then how does the Buddha explain our sense of persisting over time as enduring beings? Who does he think accumulates karma or virtue, or becomes enlightened? Furthermore, if all that exists is ever-changing and particular in nature, then how does the Buddha explain universals? How does he account for the properties of things that remain the same from one observation to the next?

Recall the interactive kind of causality that characterized particular atoms in my atom archetype versus atom instance illustration. The Buddha answers questions like the above by pointing to the systematic way one moment's interplay of phenomena gives rise to the next—a process the Buddha calls *pratītya-samutpāda*, which means roughly "interdependent arising." In Buddhism, things that we think of as enduring, including ourselves, don't preexist; they arise and are maintained through moment-by-moment causal interactions. They are expressions not of a universal unchanging basis of all things but rather of an interactive *process*.

The Buddha describes this process in terms of human beings. He talks, for example, about how our various sense faculties interact with our circumstances to cause attractions to what pleases us and repulsions from what doesn't. He talks about how these attractions and repulsions, in turn, entrench themselves to cause habitual ways of seeing and acting in the world.

Take, for example, eating a meal. Because you and I have the ability to experience tastes (an ability itself caused by prior conditions), we notice flavors in food. Because we notice these flavors, we experience the food producing a sensation that is, let's say, enjoyable. Because we enjoy the food, we look forward to the next mouthful and, beyond that, perhaps to having the meal again—and maybe again after that. Each moment thus follows from prior conditions in a never-ending stream of becoming. If any of these links change—if, for example, we don't like the meal—the consequences change as well.

According to the Buddha, our sense of self and our belief that we endure from one moment to the next arise out of our here-and-now interactions as we think of them extending over time. In other words, the names that we call ourselves—"Socrates," "Siddhartha," "Joe," "Mary"—don't identify any permanent substantial things. They point instead to sequences of here and nows strung together by an interactive process that reproduces certain features of the prior moment, including an individual's aware point of view.

FIGURE 6. Buddhist Wheel of Life. Old fresco on wall of BIA (Suan Mokkh Bangkok), Thailand. The twelve outermost segments of the wheel illustrate the twelve causal links of interdependent arising. Photo credit: Wasan Ritthawon / Shutterstock.com.

The principal aim of the Buddha's teachings is enlightenment, and his concept of "interdependent arising" (*pratītya-samutpāda*) serves this end as well. It couches the everyday, cause-and-effect behavior of existence in human terms, describing a sequence of twelve linked causal conditions that, within the framework of reincarnating human beings, circles back to the beginning like a wheel. This "Wheel of Life" of Buddhism gives a symbolic representation of our human saṃsāric predicament—how desires develop and bind us to the cycles of an ever-repeating causal chain.

Each complete cycle of twelve causal links describes one life, one enduring identity from its inception in ignorance through its coming into existence as a sentient being subject to the travails of living, aging, and dying. The first link in the chain is ignorance, followed by predispositions, consciousness, name and form, the six senses ("mind" in Buddhism is also a sense organ), contact, sensations, desires, clinging (attachment), the process of becoming, birth, and aging and death, which leads back to ignorance.

Freedom from our bondage to this cycle requires breaking this chain. To do this at the first link of ignorance, the Buddha prescribes knowledge of the nature of suffering and of release from suffering as given in the Four Noble Truths. When ignorance ceases, the first link in the chain produces no second link, which produces no third link, and so on until there are no travails of life, death, and rebirth, and one is finished with saṃsāric existence. A quest for enlightenment need not begin at the first link, however; it can begin anywhere. The links also relate in other ways besides the primary sequence; what happens in one area of life can directly affect more than one other area.[7]

It is also in terms of this interactive process that the Buddha explains universals—the properties, relations, and the like that appear to remain the same from one particular to another. Besides reproducing our own aware presence in the world, our here-and-now interactive lives reproduce innumerable other features of experience along with the names we have given them. Just as our concepts of self arise out of this process, so, too, do all the properties of the things we perceive around us.

This process also gives rise to consciousness. Unlike Hindus, the Buddha speaks of consciousness not as a first principle of nature but rather as manufactured in the moment by causal interaction in much the same way as other aspects of existence.

This does not mean that awareness is not central to the Buddha's account of reality. Like Hinduism, Buddhism has no Cartesian mind-matter split. The universe of interest is the one that we directly perceive, and its nature cannot be understood apart from the awareness that perceives it.[8] What the Buddha claims instead is that awareness from our own first-person perspective—conscious awareness (*vijñāna*)—is *particular* in nature rather than universal. Our consciousness right now depends on what we are conscious *of*. To consciously see a meal requires visual consciousness; to consciously smell a meal requires olfactory consciousness.

The Buddha emphasizes the particular nature of our conscious experience in another way as well. Recall that atom instances differ from the atom archetype by being made of parts. According to the Buddha, it is the same for each moment of experience. Not only are the things we perceive made of parts in the way an apple is made of skin, flesh, core, and seeds, but every interactive moment is ultimately composed of many momentary, extremely small "force particles" called *dhammas*. Unlike the elementary particles of matter in physics today, the dhammas are as much psychological in nature as physical; they pertain to *awareness*. They are the elementary constituents of interdependent arising examined in terms of sentient beings. Recall the prior chapter's discussion of "matter" in the context of phenomenology. The dhammas play much the same role in Buddhism as matter did in that first-person account of reality.[9]

In sum, the Buddha analyzes our awareness of a world from the perspective of our first-person views and finds us, our awareness, and all that we are aware of to be particular in nature. All that we perceive are aggregations of parts given form by a here-and-now interactive process—a process that is, furthermore, driven by desire.

Universals as Arising through Causal Interaction

I want to examine more carefully the Buddha's claim that the properties of the things we perceive around us originate in here-and-now, causal interaction, since this isn't how we usually think of them. Properties usually seem to belong to objects themselves and often seem to endure from one moment to the next. Thus we think of our cars as themselves vehicles to drive, and they continue to be vehicles to drive even after we park them for the night and go inside. Thus we think of our chairs as themselves objects to sit on, and they remain so even when no one is in the room.

Let's consider a particular situation. Instead of the proverbial tree falling in the woods with no one around to hear it, let's consider a human artifact: an alarm clock that goes off in a room when no one is around to hear it. To what extent do just the human-specific properties of the clock in fact characterize this event?[10]

Isn't it the case that when no human is in the room, human-specific properties aren't at all necessary to explain what is happening—which is little more than a lump of matter shaking the air? Are not the rules of physics entirely sufficient to describe everything that is going on? The matter itself may possess the particular alarm-clock composition of atoms that makes for the shaking, but without a person's awareness of it, this particular physical configuration has no effect as a "clock" or "alarm." It's just molecules oscillating a bit differently than usual.[11]

If a person is in the room, however, notice how all this changes. The behavior of matter by itself barely begins to tell us what is going on. If you're in the room and respond to the clock as an "alarm," then we surely need a lot more than just the rules of physics to explain why all the atoms in your body suddenly move together across the room and poke a piece of shaking matter. Our explanation could still be couched in the language of science (the language of psychology, for example, for at least part of our explanation); it just couldn't be strictly physical if it would satisfactorily account for human interests and behaviors.

This dependence of properties on who or what is involved seems to hold for everything that participates in our causally interactive universe. If no one is in your kitchen, everything there behaves as matter behaves, and that's all. When you walk into the kitchen, however, your awareness brings with it opportunities for interaction besides those of matter alone. Atoms and molecules become pots and pans, stove, refrigerator, food, and maybe eventually a cooked meal.

Likewise with properties that depend on other categories of aware beings—cats, spiders, beavers, birds, and so on. It's in the interactive involvement of members of these groups that properties particular to them become expressed. A fallen tree beside a trail in the woods can be not only a place to sit or firewood to people but also a home or meal for animals or insects—or, indeed, any number of things depending on the creatures involved. But again, the only properties actually in play at any given moment depend on who or what is there and what they're aware of. Other properties are at best latent: potential or possible properties.

Likewise, in fact, with any living thing. To whatever extent not just animals but also plants, bacteria, and so on are absent, biological properties become latent.

Likewise with matter. In whatever ways matter might be absent, material properties also become latent. (For the Buddha, the *dhammas* are always present in causal interaction.[12])

Again, this is not the way most of us think of properties in our everyday lives. Reality is typically characterized as much by these "latent" properties as by expressed ones. If the properties that made something a car or a chair were not inherent to the object itself regardless of anyone being there, then how would we know what it was when we saw it?

The Buddha does not deny the existence or importance of enduring properties like these in our lives; rather, he asks us to consider where they come from. To use my examples here, are not the enduring human-specific properties of a clock, car, or chair simply ways you and I think about and label the possibilities for here-and-now interaction? Imagine a rock formed by nature into the shape of a chair. No doubt *we* might call it a "chair," but only because it had the shape *to us* of something *we* could sit on *were we there*. Apart from how humans might interact with it, it remains only a rock, only a hunk of physical matter. Things we manufacture are no different. We may shape matter to satisfy our interests, but apart from human interests and interactions, these things, too, are just hunks of matter.

But another question now arises: What about *living beings and matter?* you might ask. If they *themselves* had no enduring distinctive qualities, then how could their participation in causal interaction be responsible for the coming and going of the properties associated with them? Surely, *something* about living beings and matter must persist from moment to moment if they are to carry opportunities for interaction with them from one occasion to another.

We're now at the crux of a problem with first-person philosophy that Buddhist themselves struggled with over the centuries.[13] No doubt first-person awareness always takes place in the here and now; yet how do we account for the continuity and persistence that we find in our world without at least *ourselves* enduring in some sense from one moment to the next?

The Buddha turned to interdependent arising and the dhammas to answer questions such as these, but he also acknowledged the difficulties of this explanation. For him, the highest truth on these matters—the truth of nirvana—is beyond conventional logic and reasoning.[14]

Enlightenment and "Presence"

In Nondual Hinduism, the goal of life is to discover one's true self, the Ātman, which is Brahman, and to live out of that discovery. The Buddha, however, teaches *anatta*, or no-self, and a cosmos without Brahman. What, then, is enlightenment in Buddhism? What does it mean to achieve nirvana?

Recall the third Noble Truth, which says that we end suffering by ending *taṇhā*, or desire. By this, the Buddha doesn't mean that nirvana is attained by ending feeling and emotion. Enlightenment means—as it does in Hinduism—the end of attachment to false conceptions of reality and to those desires that force our involvement in samsara and its sufferings.[15]

But Buddhists and Hindus generally have very different ideas of what enlightenment involves. The Hindu Upanishads approach moksha, or liberation, from the third-person view—from the standpoint of what is so for all. We achieve enlightenment by realizing that we are most fundamentally a universal—indeed, *the* universal out of which arises all things.

The Buddha, on the other hand, approaches enlightenment from the first-person view—from the standpoint of direct experience. The present moment of interacting particulars is the underlying truth, and the world of the third-person view along with the universals that describe it are manufactured out of present interactive experience. We don't seek the one universal that might underlie all the flux and change of everyday life. We seek instead to put aside universals in order to discover the present moment from which all universals arise and that holds, so we're promised, a joy beyond anything we might otherwise experience.

How might we understand this joy—the "miracle" of the present moment that Thich Nhat Hanh speaks of in my opening quote?

Consider the extent to which our conscious and unconscious interests are responsible for the properties we perceive around us. We are not just passive observers but active participants who bring possibilities for interaction to the moment that are not there without us. A confection of material ingredients becomes a piece of cake, which then becomes something to eat, which leads to a consideration about when I should eat it and, again, if it would be healthier not to eat it at all, . . . on and on. Everywhere I go, I spin a web of properties defining my reality. Quieting these desires returns the cake to something closer perhaps to just a piece of cake. The world still arises, but now it seems to me more as possibility than fact, more as a place of opportunity, art, and humor as opposed to the fixed furniture of a mind-independent universe.

Buddhists describe their goal as a kind of serenity or joy derived from experiencing life in a more mindful, insightful way.[16] Among other things, the realization of no-self, or *anatta*, brings a shift in mental focus. The I-am-a-self point of view stops being the principal driving force for how we view the world and for the properties we ascribe to reality. The world and its responsibilities remain, but our conventional concerns with identity, bodily pleasure, worldly achievement, self-esteem, and the like fade away to be replaced by compassion for others[17] and an inhabiting in and joy of the here and now—a here and now that is itself the origin of all things.[18]

Nirvana also promises greater insight into the nature of reality. In Buddhism, reality is anchored in what is present in first-person interactive awareness—in *presence*, if I might sum up this experiential basis for reality in a single word. I'll end this section with some observations about "presence" that strike me as central for understanding first-person accounts of reality.

First, notice that the presence of something in first-person experience is qualitatively different from the same thing considered apart from its presence. The properties of something that is present—what I've called "expressed" properties—come with a subjective dimension that is absent when the same thing isn't present and its properties are available only indirectly. Knowing all there is to know about a toothache, for example, is not the same as actually having a toothache. A secondhand or intellectual understanding of a dance performance or birthday party or sunset or walk down the street is not the same as actually being there. Properties as we actually encounter them in our experience are, in a certain fundamental way, different from these same properties as we find them in our thoughts about and descriptions of an experience. Direct experience reveals a dimension of properties that's stripped out when that experience becomes translated into the universals of language and logic.[19] (See "Qualia" text box.)

Second, notice that this subjective quality of what is present is in some sense ineffable. The properties we use to characterize something that is present are the *same* properties we use to characterize what isn't present. The properties by which we identify and describe an actually experienced toothache are the *same* properties we might use to identify and describe another person's toothache or a fictional one. No matter what we experience, we can't recognize, think about, or talk about it without recourse to universals—properties that remain the same from one moment to another or one person to another. And such properties render all experiences, including in-the-moment private ones, in terms of universals—what remains the same for more than just one here-and-now experience.

Third, this ineffable quality of the things that we directly experience applies most importantly to presence itself.[20] Recall from chapter 3 how the relationship between particulars and their properties is ambiguous and philosophically problematic. It's hard to understand what is left of a thing when all its shared properties are removed—when we peel away like an onion all the layers of identifying features. If presence is what would remain were we to conceptually strip our first-person view of all that was the same from one first-person view to another, then

QUALIA

In contemporary philosophy, *qualia* (singular *quale*) refer to the subjective qualities of conscious experience, or what it "is like" to personally experience something. It contrasts the experiential properties of the first-person view with their third-person counterparts. On the one hand is our actual experience of, for example, the color red— the red that you or I directly perceive in, say, a stop sign or a red evening sky. On the other hand is the third-person property of red, the red that seems to exist as a quality of the stop sign or sunset irrespective of anyone seeing it.

While it is difficult to deny that there are qualia and that they are specifically associated with our first-person view, they are philosophically controversial and at the center of efforts to understand consciousness and subjective experience. The questions of how and why there is subjective experience can be thought of as questions of how and why there are qualia—properties that seem to be subtly distinct from their third-person counterparts. What, for example, is the difference between a quale like "red" and the third-person property "red"? How might the quale "red" relate to the physical "red" (defined, for example, in terms of electromagnetic wavelength)? And perhaps most important, do qualia actually exist at all?[21]

In considering these questions, it's noteworthy that qualia are not an issue in Buddhism—at least not in the same way. If we anchor our account of reality in first-person experience, then the properties we directly experience, qualia, are our reality. Third-person properties do not exist except as labels for or pointers to qualia, to what is directly present in our experience. What Buddhists must explain is not how qualia relate to a reality that is separate from qualia but rather the apparently intersubjective nature of these "labels" as well as of the qualia to which they refer.

what might be left for the word "presence" to refer to? We appear to be left with nothing intelligible at all—nothing nameable or communicable for the word "presence" to signify.

Fourth, this in some sense ineffable presence plays a critical role in the causal interactions among aware beings. Regardless of what might exist independently of our personal awareness, we interact only with what we're in some way actually aware of.[22] If we're walking in the woods and mistake a vine for a snake, we respond to the snake that is present in our minds, not the vine that's in fact there. In other words, the accuracy of the predictions we make about the behavior of aware beings depends on understanding what is present in awareness and the difference presence makes between properties that are expressed versus ones that are merely latent.

In reflecting on the above observations about presence, you may find it somewhat perverse that I've named something, discussed it, and then called it "ineffable." Surely presence can't be ineffable, at least not entirely, if it's possible to point to it as a property that all occasions of awareness seem to share.

The difficulty posed by the concept of "presence" is that, while the *concept* may be a universal and something we can talk about, *what I want to name* with this universal is not itself a universal—at least not in the usual sense. It's more what awareness does to a universal to transform it into *this* universal instantiated here in my awareness—what transforms generic red into *this* red as I experience it. Presence is not so much itself a property as it is the experiential arena where properties enter into awareness and become the properties of this one particular moment. The problem is that I presume the *same* capability for *your* awareness, but how is that possible unless presence names what is in fact a universal?

In other words, presence seems to signify a universal that behaves in a very un-universal-like way: it turns universals into unique experiential things that aren't universal at all. (Yet how can what is truly identical from instance to instance do this?)[23]

To return to Buddhism and its goal of nirvana, the pursuit of enlightenment seems to be linked to fully grasping this at least partially ineffable "presence" in which arises the world we perceive around us.

History after the Buddha

After Siddhartha's death (dates vary from 483 to 350 B.C.E.), his followers met and collected the legacy of their time with the Buddha in what came to be called the Pāli Canon.

After a while, disputes arose over doctrine and practice, and a century or so after the first meeting, Buddhist monks broke into two main groups. One group (the *Sthaviras*, or "elders") held strictly to the teachings of the Pāli Canon of the first council. For them and those who followed them, the Buddha was principally a historical figure, and the goal of life was to become a monk, meditate, and achieve nirvana in either this or some future life. From them comes today's Theravada Buddhism of principally Sri Lanka, Thailand, Burma, Laos, and Cambodia.

The second group (the *Mahāsaṅghika*, or the "great or large community") introduced certain changes over time. For these Buddhists, the goal of life became not personal nirvana but rather a collective nirvana where all worked for the enlightenment of all sentient beings. In accord with this work, one strove to become a *bodhisattva*, one who postponed nirvana for the sake of helping others achieve enlightenment. Also, the Buddha became less a historic figure than the model of spiritual perfection. More than just a model, the Buddha became idealized—transfigured as a supermundane, perfect, eternal, omnipotent, and omniscient entity, a cosmic principle. The idealized Buddha, not the monks, was now seen as the spiritual ideal.

From this second group came Mahayana ("Great Vehicle") Buddhism, which spread north and east from India to become the most widespread form of Buddhism, the one found in Tibet, China, Korea, and Japan. Mahayana itself divided into many different groups with widely varying spiritual practices ranging from the colorful rituals of Tibetan tantra to the quiet austerity of Zen (Chán in China) to the mass appeal of Pure Land, in which a heavenly being called Amitābha Buddha (Amida Buddha in Japan) is thought to save humans through personal faith alone.

Mahayana also came with new scriptural texts supplementing (and, for some, more important than) the original Pāli Canon—texts that introduced other ways to understand reality from the first-person view. Of particular importance is the doctrine of *śūnyatā* (pronounced shoon-YAH-tuh), meaning "emptiness," specifically as applied to all that existed.[24] While the Pāli Canon interpreted *anatta*, or no-self, as the insubstantiality of the individual and the world of direct experience, the Mahayana scriptures carried the doctrine further to include the insubstantiality of absolutely *everything*, including the psychophysical force particles, the *dhammas*. Recall in this regard the unintelligibility of "presence," how it tries to identify what has no identifying qualities. One of the examples used by the early teachers of śūnyatā, the Śūnyavādins, was that of an empty vessel. They claimed that their predecessors thought of objects in the world as vessels containing nothing at all, whereas they, the Śūnyavādins, taught that even the vessels themselves were nothing at all.

Another term found in Mahayana scriptures, *tathatā*, is very close in meaning to what I've called "presence." Tathatā is typically translated as "thusness" or "suchness" and intends to refer as far as possible to the true nature of the things we perceive around us—their simple presence in awareness as an arising of properties irrespective of what those properties are. As a bare object of experience, empty in itself of any attribute or enduring nature, tathatā is also described as beyond all concepts and distinctions.

But even concepts such as these were susceptible to being thought of as referring to some actual intelligible aspect of the universe itself. In the second or third century C.E., in order to counter the tendency to turn such concepts into universal first principles of the cosmos, one of the greatest of all Buddhist teachers, Nāgārjuna, argued that every view of reality grounded in universals was empty (recall pluralism in this regard). Not only that, but even nirvana was empty; even emptiness, or śūnyatā, was empty. All that existed was the process of interdependent arising (in which even the sequential steps or links of that process were empty).[25] Later, other great Buddhists, such as Chandrakīrti (seventh century), reaffirmed these teachings against what they saw as the tendencies among Nāgārjuna's followers to turn Nāgārjuna's method of negating philosophical positions into a philosophical position itself.

Nāgārjuna and followers like Chandrakīrti form the Mādhyamika school, which is one of the two main schools of thought that grew out of the Mahayana literature. The other school is Yogācāra ("practice of yoga").[26] In response to the problem of how to explain our everyday experience of continuity and persistence, Yogācāra teaches that there is in fact one underlying reality, the reality of consciousness (*chitta*). Sometimes called the "mind-only" school, in Yogācāra, all phenomena are entirely mental creations arising from within a single consciousness or knowing process. This mind-only doctrine makes it similar in certain respects to Nondual Hinduism, a charge often made against it by the Mādhyamika and Theravada schools. Even in Yogācāra, however, enlightenment comes only to those who discover what remains of direct experience when concepts and universals are put aside.

As in Hinduism, most Buddhists are less interested in these philosophical subtleties than in what benefits might come from spiritual practice. Regardless of the Buddha's teachings about self-reliance and enlightenment, he and other awakened beings are typically revered like deities—embodiments of reality's ultimate truth and our human life goals. Indeed, the most popular of all Buddhist sects is Pure Land Buddhism, which promises an afterlife in Amitābha/Amida Buddha's heavenly Western Paradise should we have sufficient faith in his power and compassion.

Summary with Strengths and Weaknesses of This Strategy

At the heart of Buddhism is Siddhartha Gautama, the Buddha, whose life and teachings illustrate a path to enlightenment emphasizing what we can know from a first-person perspective. His Four Noble Truths sum up his teachings: life is suffering; suffering is caused by attachment, or *taṇhā*; suffering is ended by ending attachment; and the way to end attachment is via the Eightfold Path.

Philosophically, the Buddha gives what I've called a first-person account of reality. While both Hindus and Buddhists analyze reality in terms of our awareness of it and conclude that our everyday saṃsāric understanding is not the highest truth, they generally conceive of this higher truth differently. Hindus typically stress the philosophical priority of universals—what stays the same from moment to moment under all circumstances—and conceive of particulars as arising from and instantiating universals. The Buddha, on the other hand, stresses what actually presents itself in first-person experience—particulars and their causal interactions. He characterizes existence with three marks: it is impermanent (*anicca*), lacks enduring being ("no-self," or *anatta*), and is animated by suffering (*dukkha*). All that we think of as enduring or preexisting is generated in the moment by causal interaction (i.e., by *interdependent arising*) and is also a composite of parts.

It may seem counterintuitive for the world we perceive around us to arise in the moment through causal interaction, but in fact many if not most of the properties of the things we perceive do seem to depend on the here-and-now mind of the observer. They describe objects and events insofar as they might relate to us or to other humans or living beings in this or that present moment. As for other properties, those described in the prior chapter as "material," the Buddha steers a sort of first-person middle path between today's phenomenology and physics; he speaks of each moment as a composite of elementary psychophysical particles, or *dhammas*.

The goal of life for the Buddha is *nirvana*, a state of joy and peace gained by stilling those desires that drive the cycle of samsara, the world of our unexamined experience. Through enlightenment, one comes to know reality as something that arises in here-and-now interactive awareness, something that has what I've termed *presence*—a word that tries to signify what seems at least partially unsignifiable. Because aware beings interact with only what they're in some way actually aware of, presence in awareness is also key to understanding the nature of sentient interaction.

The development of Mahayana Buddhism added other concepts that emphasized and elaborated on reality as created by the mind. *Śūnyatā* refers to the emptiness or voidness of existence or of accounts of reality. *Tathatā*, a term similar in meaning to "presence," refers to the "suchness" or "thusness" of a thing grasped in its true, propertyless nature.

Mahayana's Mādhyamika and Yogācāra schools give two alternative solutions to the problem of explaining enduring beings. Mādhyamika, somewhat like Greek Sophists or today's postmodern/poststructural philosophies, points out how all philosophical justifications for anything enduring or universal are fundamentally flawed. Yogācāra, on the other hand, provides a mechanism—a "storehouse consciousness"—to account for enduring selves and universals.

Historically, Buddhism's principal philosophical difficulty has been explaining universals and enduring beings adequately when, from the first-person perspective, all things seem to arise in the moment through causal interaction. How, for example, do we explain karma and enlightenment without a self to accrue karma and become enlightened? How, too, we might add, can Buddhism adequately explain math and logic, which seem both thoroughly universal and also predictive in certain respects of what will happen?

Also notice in this context that Buddhism's first-person foundation is not without its own universals. The very process it uses to explain our experience of universals—interdependent arising—is *itself* a universal obeying enduring rules that exist throughout at least sentient experience.

Notice here as well how Buddhism's strengths and weaknesses contrast with those of Hinduism. Nondual Hinduism's concept of Brahman explains our enduring nature and the orderliness of the world but leaves us with a problem explaining particulars—how the generic idea of a car or a chair becomes *this* car or chair here or why we experience ourselves as a *particular* person (with a small *p*) as opposed to the universal Person who is Brahman. Buddhism, on the other hand, has no problem accounting for the world of particulars and here-and-now interaction but has a problem adequately explaining universals and enduring beings.

Given that the weaknesses of one strategy seem to be the strengths of the other, why shouldn't we approach reality as a combination of both strategies? We typically do approach reality this way, but this approach—my next subject—seems at least as problematic as the other two.

CHAPTER SEVEN

The Dualism of Everyday Reality

Dualist Accounts of Reality

My next subject is amalgams of first-person and third-person accounts of reality.

Given the inadequacies of the prior two accounts, it may seem as if some properly tweaked version of this third option offers the best strategy. While I do think that we combine both views of reality in our everyday practical lives, I don't think this strategy is philosophically any better than the other two. It seems invariably to result in two distinct realms of reality: one associated with the first-person view and one with the third-person view, with a relationship between them that is extremely difficult to understand.

Because they result in two realms of reality, I call these "dualist" accounts as others have. We have already seen an example of this dualism in the Cartesian distinction between the reality generated by our minds and whatever reality might exist external to our minds. On the one hand is our interior, mind-generated version of reality as presented by our first-person view. On the other hand is the true-for-all, mind-independent reality of the third-person view, which we seem to know by inference, if at all.

The dualist accounts of most interest to me at the moment, however, are those of Western theism, which have historical roots that go back well before Descartes. Instead of realms of reality based on what is and isn't of the mind, their two realms of reality divide along much the same line I've used to distinguish pre-Descartes accounts of reality—the line that divides universals from particulars. We have also already seen an example of this form of dualism in the monotheistic distinction between God and creation. God, if you recall, is typically distinguished from the world using much the same traits that distinguish universals from particulars. God is one, not many; omnipresent, not localized; incorporeal, not corporeal; unchanging and eternal, not changing or finite; inferred (or otherwise nonsensorily revealed), not empirically observed; and lawgiving, not law following.

While these associations hold for theist religions generally, they are by no means the whole story. Deities may be thought of as universals in certain respects, but they are also thought of as particular beings with whom one perhaps might communicate through ritual or prayer.

In order to better understand these non-Cartesian dualist accounts of reality, I want to take another look at our everyday idea of reality in the light of my observations over the last few chapters. Our contemporary idea of everyday reality is, if you recall, also non-Cartesian. It doesn't refer to a third-person view that's simply true for all. It refers to a third-person view that we *also* think of as something we can, at least in part, *directly experience*. As a theory of perception, this idea that we directly perceive a world that others also directly perceive is sometimes called "commonsense realism" (or "naive" or "direct" realism). Right or wrong, it's reality as it appears to us in our day-to-day direct experience.[1]

Our concept of everyday reality thus combines the *universality* of third-person accounts of reality with the *presence* of first-person accounts—but with, it seems to me, two caveats.

First, as I pointed out in my phenomenological analysis, we don't typically think of most things that make up our everyday world as absolutely true for all; rather, they are true for humans, or for cats, or for some other collection of beings. In other words, the reality of our everyday lives is not so much "universal" as it is what I've called "public." It is the same for those who share a certain way of being aware but not necessarily for everyone.

Second, recall as well the provisions made by accounts of reality for matter. Even the first-person accounts of Buddhism and phenomenology acknowledge that at least some of the properties that arise in our interactive experience seem to come from the materials (in Buddhism, the dhammas) out of which things are made. In other words, the concept of reality discussed here needs to accommodate the roles played in everyday life by not just aware beings but also matter.

Let me illustrate what I mean in this regard by pulling together these two threads of the public and the present within a single description of reality. In the process, I will also review some of my prior discussion.

Reality as What Is "Public"

When I walk down a street in my hometown, I see a variety of things, such as people and cars, buildings and trees. I see them, of course, in my own personal way, but I don't think of them as things only I can see. I think of them as things that others can see as well. The bookstore near the corner of Main and Ferry Streets is there for me to see when I walk by; it's also there for others to see when they walk by. If I look in the store window and have trouble making out the title of one of the books, I might ask someone else if she can make it out. If she can, I'll take what she says about the title to apply to me as well. There are not two book titles, one for each of us. There is only one book title for the both of us.

So we think of ourselves as living in a world that is in some sense the same for others as well as ourselves—a world that we sometimes don't perceive very clearly and can be wrong about. This "third-person view," as many have called it, is what I observed earlier to be what most of us in our everyday practical affairs usually think of as reality.

To begin with, then, reality in this everyday sense refers to what is the same for both ourselves and others irrespective of what we each may directly experience.

In order to determine if something is the same for others and not a figment of our personal imaginations, we look for things about it—properties—that are the same for others. Such properties, insofar as they have more than one instance, are "universals." They are aspects of our world that we recognize, or "re-cognize"—ones that we see again in current circumstances as being like what we and others have seen before in other circumstances. I recognize an object as a book or chair or cat because it is like other books or chairs or cats. It fits into these mental categories; it instances universal book, chair, or cat qualities. Universals like these seem to underlie all that we call "real," and it is to those universals that seem to underlie absolutely everything that philosophers turn for their first principles.

I've also discussed, however, how most of the properties that interest us in our everyday lives don't usually exist for every aware being. Sometimes, for example, ducks wander up to Main Street from the river a block away. Although they might pass by the bookstore window, I wouldn't expect creatures like these to help me in clarifying the book titles. We make distinctions as to who might perceive what, and we base our understanding of the world in part on these distinctions. Our everyday reality is made up of things that are the same for *collections* of aware individuals, not necessarily for absolutely every observer.

The word I've used to designate this intersubjective quality of the world is "public." Properties are public by virtue of being the same for collections of those who share similar ways of being aware irrespective of what members of the groups themselves may each think or perceive. In other words, public properties *depend on our minds collectively while being independent of our minds individually.* "Universals," in contrast, are defined not in terms of collections of aware individuals but rather in terms of what is the same from one *particular* to the next.

I also discussed how the physical properties of matter do not seem to depend on the awareness of any living beings but rather seem to exist for every aware being. A duck might not see books as things to read, but if a number of books were stacked in front of a duck on the sidewalk, the creature would still have to contend with them as physical objects.

In addition to matter, there are two other important cases you may have noticed that I've not yet discussed where properties also seem to be true for all. First, aware beings themselves have properties that seem true for all. In order for us to distinguish properties according to who is aware of what, aware beings must themselves be the source of sufficient characteristics to allow us to distinguish among them and their different ways of being aware (to whatever extent we are able). I'll elaborate on this exception later in this chapter.

Second is math and logic, which also seem to hold for all. I'll discuss this second exception in a later chapter.

To summarize my first distinguishing feature of everyday reality, to be real in this everyday sense is to be *public*. And to be public, something doesn't necessarily have to be universally the same for all aware individuals; it only has to be the same for collections of aware individuals or for matter. (How public properties might depend on our minds collectively while being independent of our minds individually is a question I'll take up in subsequent chapters.)

The distinction made above between something that is "public" and something that is a "universal" is critical to my discussion of everyday reality, and in order to contrast the two concepts more easily, I ask your indulgence with the word "public." Specifically, I want to replace the phrase "something that is public" with the single word: *publicity*. This word may have the unfortunate connotation of an advertising campaign, but it is also a term already used in philosophy to refer to what remains the same for a group of knowing subjects. In the philosophy of science, for example, the words "public" and "publicity" are often used when discussing the nature of evidence. Empirical evidence begins with what particular scientists observe in particular experiments done at particular times and places under particular circumstances—that is, it begins with the here-and-now data of scientists' first-person experience. But if such experimental data are to serve as evidence in support of a certain theory, they must be applicable to more than one here-and-now observation; they must have the quality of "publicity."[2]

In my context here, besides using publicity as an adjective, I also want to use the term to refer to *that which* is public. Just as we speak of properties as "universals," I also want to speak of them as "publicities." Besides simplifying my writing, this change gives me a concept that I can contrast directly with "universal"—one I can use to highlight more easily the differences between the two terms.

For one important example, notice that while a universal is something that is true for all, the concept of publicity asks who this "all" refers to. When we speak of a certain property like the "usefulness" of my desk, for example, do we mean that this property of my desk arises for absolutely all observers? What about cats and insects? Do we think the property applies to them? Surely we mean instead that the property applies to just humans, if not a subcategory thereof.

Reality as What Is "Present"

Another important difference between universals and publicities, one with direct bearing on the concept of everyday reality discussed here, is that publicities refer to *anything* that remains the same for others as well as ourselves—anything that's part of our third-person view. In other words, publicities encompass not just universals *but also particulars.* Something that is public may be a universal—for example, the property of being "red." It also may be an instance of a universal—for example, a particular patch of red, like that on the dust jacket of my dictionary. The particular red of this dust jacket is public in that it is there not just for me but also for others.

Although universals may be the quintessential publicities, when I walk down the street and look around myself, what I perceive and presume to be real are *particular* cars, *particular* trees, *particular* people having *particular* conversations. Insofar as my day-to-day perceptual experience is concerned, the public world that is real to me is made of particulars—universals as instantiated, not universals in whatever form they might have in themselves apart from their instantiations. It's this car here or that person there that is real, not the generic quality by itself of being a car or person.

This foregrounding of particulars is, if you recall, characteristic of first-person, perceptual experience, and everyday reality is explicitly the public world we know from such first-person experience. It is reality without the Cartesian divide between the mind dependent and the mind independent. To be real in this everyday sense is to participate in this world of our first-person view. What I judge to be real must not only be public (i.e., the same for a collection of individuals); it must also be in some way involved in this cosmos I know in this here-and-now, directly experienced way. It must be capable of what I've called *presence*.

Recall Buddhism's *pratītya-samutpāda*, or interdependent arising, in this context. From our first-person view, the world that exists for us is the world expressed through the here-and-now interactions of aware beings as well as, so it seems, matter. Presence is that feature of reality whereby publicities participate in here-and-now interactions—that feature whereby generic or probabilistic properties (i.e., universals, however one conceives of them) become *these particular* properties *here* or *those particular* properties *there*.[3]

For us, the direct perception of anything is absolute proof of its presence. It doesn't matter if we are hallucinating or asleep; presence has nothing to do with something being true or the same for others. It's strictly about the simple "thereness" of whatever we perceive, the arising of properties irrespective of what the properties are (what Buddhists call "suchness" or "thusness," *tathatā*).[4] If I see a person standing in front of me, the bare "thereness" or "suchness" of the experience remains a fact of my perception regardless of whether I am hallucinating. If I'm asleep and dreaming that I live in the Middle Ages, then the "presence" of that dreamscape exists for me regardless of where I would find myself when I woke up. If I'm walking along a path in the woods and mistake a vine for a snake, the object I see is no less present to me for my error. In direct perception, regardless of whether I recognize an object correctly or whether I have the right properties, the presence of the interaction remains. (And it's to properties actually present to me that I respond regardless of whether I have them right.)

The presence that defines this second aspect of everyday reality is not confined to just direct perception, however. I don't typically think of everyday realities as things that go away when I'm not looking. When I think of such things, part of what they are for me is their *potential* for presence, their *latent* properties. When I park my car in the parking lot in the evening and go inside, I still think of my car as a real car that I could drive if I wanted. In everyday life, we don't typically reserve the word "real" for only what is present right now in our perceptual experience (as we might if discussing Buddhism). We also use it for what we think of as capable of presence, what would be there if we or someone else were to go take a look.[5]

In other words, the presence that combines with publicity in this concept of everyday reality is the *capability* for presence. And in this case, we can be wrong about presence. Since the capability for presence is self-evident only in what we directly experience, in all other situations, it must be figured out in conjunction with other properties.[6]

Finally, the capability for presence that defines everyday reality in my context here applies to any interactive involvement in the universe at all. The properties that make

COUNTERFACTUAL CONDITIONALS

In considering how reality might include capabilities for presence, consider our everyday use of "counterfactual conditionals"—statements about reality that are contrary to fact yet seem to be true because of the conditional way in which they are stated. For example, consider the sentence "If I had bought milk when I went to the store yesterday, I wouldn't have run out of milk this morning." Such a sentence describes a causal sequence that seems true even though it never happened; it seems to describe an alternative path that reality might have taken had something occurred that did not.

Counterfactuals like this are commonplace in our thinking and appear to tell us much about our world. But they pose a philosophical problem insofar as the causal behavior of reality is based on universal rules of nature. Why do they seem true when they don't correspond to what actually occurs? Why should they be so instructive when they describe situations that are contrary to what we know happens or exists?

Consider, however, a concept of reality where the causal nature of what we call "real" doesn't come from simply universal rules but rather from rules with various capabilities for presence. In that case, counterfactual conditionals give the alternative possibilities that in fact characterize reality at a given time.[7]

up our universe come from other sources besides just human beings. A certain photon's minute contribution to the force fields of our universe is enough to establish its presence. What we conventionally think of as "real" applies to anything capable of participating in the causal interactions of our universe regardless if that is a human or ant or virus or inert matter.[8]

In sum, the concept of everyday reality I'm interested in here is an amalgam of publicity and presence. Regardless of our philosophical views about reality, we seem to go about our day-to-day practical affairs as if we lived in a world that was both public and capable of presence.[9] In our everyday lives, we are all "commonsense realists," as philosophers put it; on this much, at least, philosophers seem to agree.[10] It's rather when we probe deeper that these elements of our everyday reality become controversial. It's then that we come to a triple fork in our path where we seem forced to choose before going farther. We can take the third-person route and anchor our accounts of reality in what is public (and more broadly true for all), as does the nondual Hindu school as well as much of Western philosophy. Turning the other way, we can take the first-person route and anchor our accounts of reality in what is present for us as do most Buddhist schools and certain more recent phenomenological/postmodern/poststructural/existential trends. Or as we will see next, we can proceed straight ahead with a dualist view of reality, as we tend to do in Western monotheistic religions. (This is not to deny other efforts to reconcile the first-person and third-person views, though none of them seems to escape philosophical problems associated with the two views.)

Phusis versus *Techne*: The Natural versus the Artificial

Certainly the bulk of the properties we humans use to describe our world come from us and reflect human-specific interests. Nearly all the things I see around me in my office have properties that I and other people recognize but that I don't think of as characterizing the realities of cats, beetles, bats, and other nonhuman creatures. For their part, these other creatures no doubt perceive a world characterized by properties that depend on their own perceptual apparatus and interests. In this respect, most of the properties that describe the realities of aware beings—human or otherwise—come from the way different groups of beings are aware.

One of the main exceptions I've mentioned is matter, which presents itself to us as having properties that seem to come from matter itself. But another important exception I've pointed out is aware beings. Aware beings also present themselves to us in our everyday lives as having distinctive traits for which they, not us, are the source. Moreover, as the apparent source of at least some of their own properties, they present themselves to us as having in some sense an independent existence—as being self-existent in some way (although whether such beings are in fact self-existent has been much debated[11]).

The distinction between things that are self-existent like this and things that are produced by such self-existent things is an ancient and important one. I mentioned that the beginning of Greek philosophy introduced new concepts, one being that all things were parts of a whole, a *kosmos* that was characterized by universal properties. But also new was the idea that certain things in the universe had a "nature" to them, one based on logical principles, a *phusis* (from which we get our word "physics"). Early Greek philosophers called themselves *phusikoi*, students of phusis. Phusis means the inherent nature of those things of the world that are not made—that grow, that have their own being, that can be treated as things in themselves and the source of other things.[12]

The distinction the Greeks made here was between phusis and *techne*, between the natural and the artificial, the self-existing and what comes from another. On the one hand are people, animals, trees, and for these ancient people also things like the sun, moon, and stars. On the other hand are houses, carts, pots, and ploughs. Phusis distinguishes things that are the source of their own existence as well as the existence of other things. Techne, on the other hand, identifies *artifacts*, or things that come from those things with a phusis.[13]

Phusis didn't refer to just the nature of individual things. As in today's use of the word "nature," it also refers to nature in general, as when we talk about the "nature of the universe" or the "laws of nature." One of the founding assumptions of Western philosophy was the idea that the cosmos itself had a nature, that logical principles inherent to the cosmos ran through all things and knit them together into a whole. This nature was thought to be evident in the material out of which everything was made. Thus Thales, so we are told, believed that everything was made out of water, while Heraclitus thought that it was made out of fire. Democritus (c. 460–370 B.C.E.), for his part, believed that everything was made out of small atom-like particles. In all these cases, the logical nature of the universe as a whole determined, at least to some extent, how everything in it behaved.

This distinction between the natural and the artificial remains central to science today. To investigate the nature of an atom's nucleus, for example, is to seek to understand its phusis—those basic properties (perhaps its quantum mechanical structure) in terms of which other properties such as size, composition, density, electrical charge, or reactions to impact can be explained. Assumed is the idea that physical phenomena obey certain universal rules in terms of which we might understand the properties we observe in all things made of matter.

The same is true of living organisms. In biology, we study living beings as a separate category of things that share certain characteristics not found in inert matter (e.g., RNA/DNA) that, once known, explain, for example, the similarities and differences we find among living creatures. Likewise for human beings. We study the human mind in its own right as having its own distinctive rationale (perhaps its cognitive structure) on the basis of which our different, particularly human reactions to circumstances might be understood.

You may have noticed that I've given two different lists for what presents itself to us as the source of its own properties. For Greek philosophers and, I expect, most of us today, this list is made up of living beings and matter. Earlier, however, it was "aware" beings, not "living" beings, that apart from matter appeared to have the capability of presenting their own properties.

Perhaps a question to ask in this regard is whether these two lists are not in fact the same. In other words, might it not be awareness in some form or another that provides the capability of presenting properties other than the physical properties of matter—the capability of interacting in ways that cannot be explained by physics alone?

Individual and Collective Agency

Phusis distinguishes the natural from the artificial, the self-existing from what comes from another. It identifies those things—typically living beings and matter—that we think of as the source of at least some of their own properties as well as the properties of other things.

To speak about something as the source of other things is to speak about *agency*, or where things originate in our universe. In this section, I want to distinguish between two different kinds of agency. To do this, I'll start not with the concept of agency itself but rather with the related concept of causality.

My chapter 3 contrast between atom instances and atom archetype distinguished two kinds of causal relationships, two places we can look for the cause of an atom's behavior. First, we can trace an atom's behavior to the properties shared by all atoms (the atom "archetype" as I called it). In this case, we find an atom's behavior determined by *collective* properties, by the *collective* rules all atoms follow. Second, we can trace an atom's behavior to its *individual* circumstances and what's affecting it at a certain time and place. This is the cause-and-effect, one-thing-happens-which-causes-another-thing-to-happen variety of causality (causation as we usually think of it) that we find in the interactions of *individuals*. On the one hand, behavior is determined by enduring properties, by factors that do not change from one situation to another. On the other hand, behavior is determined by factors that are specific to individual circumstances and do change from instance to instance, factors that govern the way enduring properties are or are not expressed in a particular situation.

These two kinds of causality may give us origins in some sense, but they don't seem to give us *agency*. They don't tell us what's behind the causality—what it is that links an archetype to its instances or one here and now to another. In other words, they don't tell us the *cause of the causality*. They don't explain how causality might be related to things with a nature, a phusis.[14]

While causality may not give us agency directly, it does seem to give us clues about agency. To speak of agency is typically to speak of a cause that is not itself caused by anything else or a cause that is in some sense intentional or self-willed.[15]

Especially important in my context here, the distinction I've made between two kinds of causality would seem to apply as well to the way we think about agency in our day-to-day lives. We may not understand how things with a nature, a phusis, are the origin of events and circumstances (or even if they are in fact the ultimate origin); nevertheless, as an everyday practical matter, we trace certain things to individuals and other things to collections of individuals.

Consider first *individual agency* (as I'll call it here): how we—and our legal system—trace some of what happens in life to individuals. If I purposely throw a rock through a glass window, then I as an individual am usually held responsible. When we build a house, it's individual people that lay each brick or hammer each nail. Likewise with other living beings. When a hawk hunts for food, it's the individual hawk that finds and kills its prey. When an ant gathers up a breadcrumb from a kitchen counter, it's the individual ant that hefts the object and negotiates its path home.[16]

But individual agency doesn't account for many other things. We need another kind of agency to explain why groups of things have certain properties in common— why one ant looks and behaves much like other ants, one stalk of wheat like other stalks, one biological cell like other cells, or one atom like other atoms. In other words, individual agency does not explain universals. (Buddhists, for example, turn to a universal process of causal interaction for such an explanation.)

I've already pointed out how universals are not so much true for all as true for collections of living beings or for matter. More to the point of this section, notice that it is to these collections of living beings or to matter that we turn to *understand* universals; it's to *collective agency* rather than individual agency that we look for the origins of universals.[17]

Suppose, for example, that I want to understand why the teacup on my desk is shaped as it is. I'm not at the moment interested in individual agency—who in particular made the cup or how it came to be in my house or why it's now sitting on my desk. Instead, I want to understand this teacup as an instance of all teacups— why teacups have the properties they do and what makes this teacup a "teacup" and the same as other teacups.

Notice that we cannot find an answer to these questions by studying just teacups. Teacups don't make teacups; people make them. To understand why a teacup is as it is, I need to study human beings and how this particular physical form is useful to people for drinking liquids and why humans like to drink tea out of objects shaped like this.

But studying human beings will not be enough. Recall the differences between human-specific properties and material properties—how the cup seems to combine both into the idea of "teacup." Not only does it have properties that trace back to human beings, but it also has physical properties that trace back to the material out of which it is made, that describe why it is solid, that describe why this shape and material composition holds liquids, and so on.

In other words, if we are to understand why a teacup is as it is, we need to examine at least two different collections of things: physical things and human beings.[18]

Likewise with other things in our lives. If I want to understand anthills, I could learn a certain amount by studying just anthills and their material construction, but I'd learn much more by also studying ants. If I want to understand the properties of water, I need to study only matter, but if I want to understand the properties of cytoplasm, I should study cells as well as matter. If I want to understand beavers and birds and things made by them, I should study not just matter and cells but also multicellular organisms.

Universals, then (except for those of logic and math), appear in everyday life not only to be associated with but also in some sense to *originate with* collections of living beings or matter. That is, they appear to originate with collections of things with a phusis, things that are the source of at least some of their own properties as well as the properties of other things. We might question whether certain collections of things in fact originate their own and other properties, but such a question is again one about where these properties come from. Do living beings have properties that in fact originate with these living beings, or might not these properties instead originate in part or entirely with matter, or with the universe as a whole (the collection or collections of all things), or with just the human mind (our collective human way of being aware)? Regardless of one's views on this subject, the question is itself one of collective agency, of where properties originate.[19]

Before moving on to how collective agency might relate to the concept of deity, I want to return to the concept of "publicity" and how it differs from that of "universal."

Part of what the term publicity gives us, it seems to me, is a more precisely defined universal—one delimited not *just* by a class of instances but *also* by the collection(s) of living beings and/or matter to which the universal traces (which might in some cases arguably be all living beings and matter). The publicity "teacup" is *not only* a class encompassing all teacups; it is *also* a class that in some sense *originates* with human beings as well as matter.

Moreover, if we do not limit the collective sources for publicities to just those distributed in space but include those distributed in *time* as well, then such sources

UNIVERSALS AND THE "GRUE" PROBLEM

Our principal tool for finding universals in the world around us is inductive logic. We generalize from a sampling of instances to what seems so in all cases. If every crow we've ever seen has been black, then it's quite likely that the next crow we see will also be black and that black is a universal property of crows.

This way of arriving at a universal from a set of data may seem fairly straightforward. But in the mid-twentieth century, the philosopher Nelson Goodman pointed out that our formal understanding of inductive logic was incomplete. Something else seemed required of a universal besides it simply fitting the data at hand. Specifically, from a given set of data, formal logic seems to allow us to infer properties—or "predicates" (Goodman's preferred term)—that appear to be absurd, ones that we would intuitively never consider in everyday life. Goodman illustrates this problem with specific examples, the best known of which is the "grue" problem.

In the grue problem, Goodman invents a preposterous but logically permissible predicate called "grue." Things are grue if their color is always green when examined before a certain time but blue in all other cases (i.e., if they haven't been examined before the specified time or if they're examined after that time). He then asks us to consider this property for emeralds with the specified time in the future.

For example, if the specified time is midnight tonight and the emeralds are classic emerald green, then after midnight they will appear green. However, if they are colored grue, they will appear blue. From the perspective of traditional formal logic, grue describes all the emeralds we've ever seen just as well as green does. Intuitively, however, we would not presume that emeralds are grue. But why not? What parameter is formal logic missing that in everyday life distinguishes predicates that simply describe what we've observed from predicates able to tell us what will happen beyond our observations?

Clearly, it's not adequate in inductive logic to base predication simply on traits that fit all observed instances; something else is required.

> The analysis in this book suggests that the predicates we intuitively use to describe our world are based not simply on universals but rather on universals that have in fact an additional parameter, a parameter that specifies the collective agents to which universals trace.[20]

can also be *single* aware beings considered as collections of sequential moments of awareness. In other words, the term publicity can apply to "private" properties—our distinctive characteristics of thought or behavior, such as, for me, my personal memories of the books and souvenirs on my bookcase that are enduring parts of my experience of them.

In this fashion, this added parameter of the collective source(s) provides what seems to me to be a more broadly applicable and accurate way to talk about universals, one that can also be used in contexts like Buddhism that trace properties to sequential here-and-now sentient interactions. The temporal context for publicities also offers another way to understand personal identity and our sense of having enduring natures. (I'll return to these subjects later.)

This distinction between universals and publicities is one of the most important in the book, so let me repeat and elaborate on it a little.

A universal is what is the same from one particular to another. A publicity is a universal *or particular* to which a parameter is added, one that specifies the collection(s) of entities to which the universal or particular traces (living beings, material simples,[21] etc.). The Canada-United States border traces to human beings; the property of being an anthill traces, at least in part, to ants; gravity traces to matter.[22]

In the context of aware beings, while a universal is, again, what is the same from one particular to another, publicities can be thought of as what is the same *from one observer to another*, but with a caveat: an "observer" in this case doesn't just "observe." Observers are actors in the universe, participants in collections that originate publicities.[23]

The entities in an originating collection need not appear well delineated to other entities. They can present themselves as only probabilities (as in the electron shells of atoms). What is important is the collection as itself a distinct feature of the universe apart from its members—the collection as a "system," "network," or "field" (as in the case of matter).

Chapters 10 and 11 will further discuss the nature of these collections.

Summary

What I call "third-person" accounts of reality are those anchored in universal first principles, while "first-person" accounts are those anchored in what is present in our first-person views. "Dualist" accounts of reality are those that try to accommodate both approaches; however, they seem invariably to result in two realms of reality with a relationship between them that's very hard to understand.

The dualist accounts of reality that I want to focus on in the next chapter have their origin before the Enlightenment and the Cartesian gap between inner mind and outer matter. Our contemporary idea of everyday reality is similar, and it is in this context that I'll examine theist accounts of reality.

The idea of everyday reality I'm interested in has two aspects. Drawing from third-person accounts, what we call "real" is *public* (adjective, and "publicity," the noun form): it's the same for others as well as ourselves. To be public, a thing need not be the same for all, just the same for those who share a similar way of being aware or for matter. Drawing from first-person accounts, what is real is *capable of presence*: it can participate in the here-and-now interactions of our universe. Anything that so participates is real in this sense, including plants and matter.

The advent of Greek philosophy brought with it a distinction between the "natural" and the "artificial," *phusis* and *techne*, that still informs philosophy and science today. On the one hand are things that are the source of their own existence as well as the existence of other things (in everyday life, typically living beings and matter); on the other hand are the consequences or artifacts we trace back to them (houses, anthills, planets).

To consider living beings and matter (or anything else) to be the sources of what exists in the universe is to associate them with *agency*, which can be distinguished into two kinds. *Individual agency* is the generation of what exists insofar as it originates with individuals; *collective agency* is the generation of what exists by collections of individuals.

The concept of publicity as I use it here presumes not only an additional parameter for universals, that of their collective source(s), but also that a collective source can be a sequence of individual here-and-now occasions of awareness over time, as is the case for a single aware being.

CHAPTER EIGHT

Western Theism and the Dualist View

Hear, O Israel: the Lord is our God, the Lord is One. Blessed is the Name of His glorious kingdom for all eternity. You shall love the Lord your God with all your heart, with all your soul and with all your might.

> *Beginning of the Shema (Deut. 6:4–9 with added blessing),*
> *the central and most important prayer in Judaism*

Our Father in heaven, hallowed be your name. Your kingdom come. Your will be done, on earth as it is in heaven.

> *Beginning of the Lord's Prayer (Matt. 6:9–13; var. Luke 11:2–4), ascribed*
> *to Jesus and the best-known and most repeated Christian prayer*

In the Name of Allah, the Compassionate, the Merciful. Praise be to Allah, the Lord of the Worlds, the Compassionate, the Merciful, Master of the Day of Judgment. Only You do we worship, and only You do we implore for help.

> *Beginning of the Qur'an's brief opening chapter, which is considered the*
> *quintessence of the Qur'an and is a part of Islam's prescribed daily prayers*

Everyday Dualism versus a Higher Truth

The concept of everyday reality I've been discussing combines publicity and presence. It refers to what is the same for others (what is "public") together with what can be involved in the interactions of living beings or matter (what is capable of "presence").

Central to this characterization of reality is the distinction between individuals and collections of individuals. To be present is to play a part in individual, here-and-now interaction. To be public is to be the same over a collection of such here-and-now interactions.

This individual-collective distinction has also become central to my discussion in other ways. First-person accounts of reality are based on the first-person view, the view of individuals; third-person accounts are based on what remains the same over collections of such first-person views (and matter). The contrast discussed in the previous chapter between individual and collective agency is also based on this distinction. In certain respects, what exists traces back to individuals in themselves and their here-and-now causal interactions; in other respects, what exists traces back to collections of individuals taken together.

We now face the puzzle of how to understand this division of duties between the individual and the collective. We know this puzzle perhaps best in the social roles we play as both individuals and members of groups—how we balance our individual and collective interests in our families, communities, and other interpersonal relationships. Our understanding of ourselves in these roles governs our ethics, morality, and values, our political and legal systems, our educational institutions, our choices of what to teach and learn, and the roles we seek to play in society and the world.

The main questions before us now, however, concern accounts of reality. How might our public world have properties that seem independent of us as individuals yet not independent of us as members of groups? What, moreover, does this double role we play as both individuals and members of groups tell us about human nature and perhaps also the nature of aware beings in general?

Despite their different accounts of reality, Nondual Hindus and Buddhists agree in at least one respect. For them, our everyday understanding of ourselves and our relation to the world is an error. In our day-to-day experience, it may seem to us that the public properties of the world around us are mind independent—that is, they exist irrespective of our awareness of them. But in these religions, this everyday experience of reality is *samsara*—what is true for the unenlightened. It's not the highest truth, the realization of which reveals these public properties to depend in some sense on us—our minds and the nature of our awareness.

This chapter describes yet another way to understand this self-world relationship. Rather than there being a higher truth that overcomes the dualism of our everyday self-versus-world experience, this approach keeps this dualism intact. In this case, our sense of being enduring individuals living in a separate-from-us public world is *not* an error. In the final analysis, humans are "souls," or "persons," or "egos," or "selves," and we are fundamentally distinct from the other people and things we see around us. As I'll discuss next, we are beings of individual agency only; collective agency lies elsewhere. The public properties of our shared experience do not, in the final analysis, come from us. They are perhaps expressions of the nature of the universe or of evolutionary happenstance, or, historically more often, we trace them to one or more deities.

The Western theist religions to be discussed in this chapter cover a great variety of traditions and beliefs. They encompass the three Abrahamic monotheist traditions (Judaism, Christianity, and Islam) along with their polytheist ancestors. Nevertheless, as in previous chapters, my focus is very narrow. In this case, I want to lay out a certain rationale I find underlying the concept of deity while fleshing out what I've called dualist accounts of reality.

Polytheism

The philosopher William James commented that "polytheism . . . has always been the real religion of common people."[1] Certainly the belief in many gods has been humanity's standard fare for most of its recorded history, and the transition to monotheism came neither easily nor at all completely. Polytheist gods have lived on in monotheism as angels, demons, and jinn, as well as the saints, prophets, saviors, and other personages to whom people appeal in worship and prayer.

Insofar as the philosophy of theism is concerned, I've discussed how much of what is important to us in our everyday world originates with collective rather than individual agency. The shared properties and the orderliness of nature do not come from us or any other creatures or things as individuals. Furthermore, given the central importance of collective agency in its various forms, we might expect people historically to have singled out and named those agents of the collective that were thought to be the most important and central to their ways of life.

In fact, polytheist cultures have at one time or another associated almost every significant aspect of collective experience with deities. There have been gods or goddesses of birth, death, earth, sky, fertility, planting, hunting, particular rooms of a house, stoves, cities, mountains, valleys, waterfalls, wisdom, war—to name a very few.

Among the possibilities, the particular aspects of life expressed as deities tend to reflect the cultural context.[2] Hunting societies typically have not only deities associated with ancestors and the sky, which are common to many societies, but also ones associated with animals—ancestors or heroes in the form of animals or rulers of animals. On the other hand, both planting and pastoral societies emphasize deities of earth fertility, vegetation, atmosphere, and weather—gods of sun or storms, for example. Besides deities like these, planting societies also tend to have spiritual beings directly associated with crops, such as a goddess of grain or harvest. Pastoral societies, in contrast, typically have deities associated with herding or types of animals: a cow goddess or a god of shepherds. More complex, stratified societies tend to come with pantheons of gods and goddesses typically organized hierarchically according to power and importance.[3]

The deities of Mesopotamia prior to the advent of Jewish monotheism and Greek philosophy were similar. They were the sources of order and disorder in the world—what explained to people the world's behavior. *Nanna*, a Sumerian moon god, was not just the moon but the moon's power to affect human life. He was the principal agent of the night sky, lighting the land for hunting and herding and measuring time by his changing appearance (in English, the word "moon" has the same linguistic root as "month" and "measure"). *Inanna* (*Ishtar* in Akkadian) in one of her many forms was the power of the sky and spring thundershowers to bring green pastures.[4]

Important also was the role these deities played in giving human life meaning and value. They identified agents of nature in the context of a culture's central concerns and provided concepts that shaped people's activities and nurtured collective values. To worship an earth fertility goddess like Inanna was to habitually reaffirm one's tie to the land, to pastures or crops, to the seasonal cycles of plant life, and to the power that gives nature its order and disorder in these matters. Worshipping the same deities collectively and over time brought generations of people under the same conceptual roof, the same worldview. By making offerings to a goddess of earth fertility or to a sun god, members of a community affirmed their relationship to that deity and the part each person played in a cosmic drama much larger than themselves.[5]

Two observations seem especially relevant for understanding polytheism as an account of reality in my context here. First is the fundamental role deities play in nature's order. Contemporary Hindus and Buddhists also believe in deities, but in early Mesopotamian polytheism, there was no underlying first principle like Brahman or a universal process like interdependent arising that lay beyond and was more basic than deities for explaining how the universe behaved. There was no enlightenment experience that revealed a level of reality or agency beyond the gods and goddesses. Deities were final realities of nature.

Second is the difference between collective agency as a deity and how we usually think of collective agency today. We may today, for example, think of all cows as having a shared nature, but we don't usually consider that nature to be an autonomous agent of the universe. For us, cows obey physical and biological rules shared by all living creatures. To be a polytheist deity, however, a certain collective aspect of community life must be thought of as *itself* in some sense a final agent of the public world—something with its *own* self-existent nature (a *phusis*).

I've already discussed two major philosophical milestones that historically undermined such polytheist accounts of reality. First, the advent of philosophy itself promoted universal first principles over local deities as the ultimate basis for nature's order. Second, the Enlightenment brought the Cartesian gap, which interiorized the entire landscape of first-person experience—especially all that we might think of as life, awareness, and mind.

As I mentioned in chapter 2, this second milestone has been especially troublesome to religions in modern times. Recall how much of what we perceive around us is made of properties that are human specific. They depend on our shared ways of being aware and originate within our collective psyche and interactions. For pre-Descartes polytheists, there was no interior mind versus exterior matter division as we know it today. The spiritual beings who put faces on the forces of our collective experience were free to roam the times and spaces and relationships among individuals. The Enlightenment, however, confined these agents of nature to our individual minds and brains. Only matter and physical forces seemed left to dwell in the spaces among us.[6]

In sum, the deities of the prephilosophy Near East are those agents of the collective world that, among other things, explained the world's order and reoccurring properties. Agents of the collective remain today but are not usually thought of as final agents of nature, even in a limited way. Furthermore, apart from matter, they are typically relegated to our minds and a realm of experience interior to our sense organs.

From Polytheism to Monotheism

The three Abrahamic religions trace historically to the second-millennium B.C.E. Hebrews, a loose-knit collection of Hebrew-speaking people living in and to the west of Mesopotamia, the land around the Tigris and Euphrates Rivers in what is now Iraq. For the most part, they were probably nomadic cattle and sheep herders, and like others at the time, multiple deities figured prominently in their conception of reality.

The concept of one God apparently developed slowly. The books that make up the Jewish scriptures, the Tanakh (roughly the same books that make up the Christian Old Testament), are a collection of many writings that, among other things, tell the story of the Jewish people up through most of the first millennium B.C.E. The first five of these books, the Torah, include some of the earliest writings and are especially revered; yet one finds in the early texts little evidence of a belief that there exists only one God.[7] Instead, they tell the story of the relationship of the Hebrew-speaking people to one particular god, *Yahweh*, an especially powerful creator god probably modeled at least in part on the Canaanite creator god, *El*. For example, the Torah tells of Yahweh giving ten commandments to the Hebrew people, the first of which states, "I, Yahweh, am your God who brought you out of the land of Egypt, the house of bondage: You shall have no other gods besides Me" (Ex. 20:1–3). This commandment does not tell the Hebrews that there is no other god; nor does it command them to believe that there is only one God. It tells them, instead, to worship only one god, Yahweh, as their tribal deity to the exclusion of all other deities.

Even though Yahweh apparently began as just a tribal god, one of his remarkable traits from early on was the extent to which he seems to have been involved in all aspects of community life. The Tanakh doesn't portray Yahweh as a limited god with a circumscribed province of authority; his authority over the Hebrews was universal. Unlike gods of neighboring cultures, he claims complete jurisdiction over all areas of life.[8]

In Yahweh, then, the Hebrews saw themselves linked together by one overarching source for their reality as opposed to many. Yahweh was the one all-powerful, collective agent for the Hebrew-speaking community.

There are many indications in scripture and the Hebrew language that Yahweh as the creator deity was thought of as an archetype-like basis for human nature. Perhaps the most obvious are the Genesis passages where God creates Adam, the first human: "God created man in His image, in the image of God He created him" (Gen. 1:27), and then "The Lord God . . . blew into his nostrils the breath of life, and man became a living being" (Gen. 2:7). Since the Hebrew word for "breath" is also the word for "soul" and "life spirit," this passage suggests that God's own soul or life spirit is breathed into matter and becomes our human souls and life spirits.[9]

But scriptural references like these give only half the story. The Hebrew texts also emphasize the *separation* between humankind and Yahweh. God is not the Brahman of Nondual Hinduism. As I discussed in chapter 4, Genesis depicts God not as a self who evolves into creation but rather as a being who creates the world from the outside much as an artisan would. The final product is a world with a nature that is distinct in certain basic ways from God's own nature.

This distinction between God and creation runs throughout the Tanakh. After the account of creation, Hebrew scriptures do not admonish the Israelites to discover God as identical with the human soul as do the Upanishads. The human-God dialectic is not presumed to be one between the person and that person's own concealed true self. It is instead one that takes place across a divide between two sorts of beings.[10] Religious injunctions are about keeping a covenant with a deity that is distinct in its nature from humans. If the Israelites keep their covenants with God, they will be rewarded; if not, they will be punished.

Thus Yahweh appears in the Tanakh as *both* archetypal *and* separate. When, for example, prophets talk to God, the conversation often seems like one between two people, but with one of them being the all-powerful, collective agent of Hebrew life:

Where[, O Lord,] can I escape from Your spirit?
Where can I flee from Your presence?
If I ascend to heaven, You are there;
   if I descend to Sheol [the abode of the dead], You are there too.
If I take wing with the dawn
   to come to rest on the western horizon,
   even there Your hand will be guiding me,
   Your right hand will be holding me fast.

. . .

It is You who created my conscience;
   You fashioned me in my mother's womb. (Ps. 139:7–9, 13)

Over the following centuries, Yahweh's dominion extended even further, and he eventually became the one overarching agent of the entire universe. The book of Isaiah in the Tanakh is thought to have not one but three authors separated by several centuries. By the time of the second author, perhaps the sixth century B.C.E. (though dates range from seventh to fifth century), God has become the one and only final power and authority over not just the Israelites but all people and all things.[11]

As the one all-powerful agent of the universe as a whole, Yahweh continues to be Brahman-like on the one hand yet separate from creation on the other. On the one hand, Yahweh is singular, omnipresent, eternal, incorporeal, and lawgiving— the one archetypal, collective agent whereby the universe is as it is. On the other hand, Yahweh remains an agent apart from creation and human beings, a being whose nature is fundamentally different from our own.

I'd like to end this section by highlighting a couple of things about the depiction of Yahweh in the Tanakh.

First, notice that this "separation" of God from creation is one of *agency*. It isn't one of time or space, since God is present everywhere at all times.

Second, notice that the Tanakh portrays Yahweh's agency in the universe in a way that is somewhat different from the way it is portrayed by most modern theologians. Since at least Augustine (354–430 C.E.), Abrahamic writers have usually considered God's supreme and pervasive agency to be of a kind that raised difficult questions about the nature of human agency and, specifically, what we think of as free will—our capacity to choose or act of our own accord as opposed to all that we think and do being determined by factors outside our control. If God is indeed all powerful and all knowing and if divine will and foreknowledge predestine the course of all events, then, as the argument goes, how could any choice we make be in fact our own?

This idea that divine agency might contradict humans as themselves agents does not come from scripture, however. While predestination and free will are both scriptural subjects, the idea that they contradict each other is not. I know of no passage in the Tanakh or Christian New Testament or Qur'an that wonders how God can predestine the course of events and humans can be agents as well. *Both* are the case (and often paradoxically so).[12] A common metaphor in the Tanakh for Yahweh's power is that of a king who rules over his subjects as would an earthly king.[13] God and humans are *both* agencies, *even though* it is God who establishes the rules that determine the ultimate course of events. As in our day-to-day social affairs, we don't experience collective agency as necessarily precluding individual agency. Both exist as aspects of our reality.

Our problem of understanding free will (the one Augustine and so many others have pondered) comes not from scripture or everyday experience but from our way of looking at the cosmos that came with the advent of Greek philosophy. Indeed, from this vantage point, the problem of human free will arises whether or not one believes in God. If first principles, logic, and circumstance (and, today, the laws of physics, DNA, cultural memes, etc.) entirely govern all that we do, then no room remains here either for voluntary choice. (See also the "Free Will, Evil, and Accounts of Reality" text box on the following page.)

Philosophical Influences

Not surprisingly, we find Western philosophy appearing historically at about the same time as Isaiah's monotheism. Both Western philosophy and monotheism rely on shared ideas. Both replace the ultimate authority of local deities with a single, universal authority. Western philosophy conceives of this authority as the logical behavior of universal principles; monotheism conceives of it as the one God of all creation. In both cases, the feudal politics of the polytheistic heavens are overthrown by a higher universal power, one that, at least initially, did not seem to preclude the free will of people or the existence of lesser gods and goddesses.

Greek philosophers, especially Plato and Aristotle, greatly affected the development of Western monotheism, but their influence encouraged theology in a different direction than the dominant philosophical trends in India at the time.

FREE WILL, EVIL, AND ACCOUNTS OF REALITY

Within monotheist religions, the question of whether we have free will has never been satisfactorily resolved. What is often overlooked, however, is that this question of free will is not an issue in all accounts of reality. It doesn't arise within Nondual Hinduism or Buddhism (at least not in the same way) because their accounts of reality do not view human volition as an agency unto itself—something apart from Brahman or from interdependent arising. This doesn't mean that Hinduism and Buddhism don't have philosophical problems related to human nature; indeed, they have their own different problems that dualist accounts avoid.[14] Rather, it means that free will poses a problem only when individual agency and collective agency don't have a common source; it's a problem only in dualist accounts of reality.

Notice, too, that the dualist strategy for understanding reality is also responsible for the theological problem of evil. If God is in fact all powerful and all knowing, then God should be responsible for the existence of evil, which contradicts the idea of God as compassionate and merciful. This problem also doesn't arise in the same way in these Eastern religions, since, again, there is no agency apart from ourselves or from interdependent arising to cause whatever evil there is in the world.

Because these problems of free will and evil are particular to—are baked into, so to speak—dualist strategies for understanding reality, it seems pointless to pursue answers within dualist accounts by themselves. It would seem more fruitful to look within a larger context encompassing first-person and third-person accounts. What, for example, divides the first-person and third-person views and, in so doing, produces two contrasting foundations for our accounts of reality? How do these two views of reality relate to each other across this divide? What might the problems of free will and evil tell us about these two views of reality and their relationship to each other?

For Plato, this East-West difference is especially obvious in his characterization of the human soul. The "soul" (*psuchè*, our word "psyche") is his explanation for how universals (i.e., the "forms") become instantiated in particulars. Souls are those eternal and indestructible agents of the cosmos that fashion universals into the particular objects of sense experience.[15] Thus, for example, the universal and perfect forms of Virtue and Beauty come to be expressed, however imperfectly, in the character and works of individual people. Our task is to discover and instantiate these universals (of which the most fundamental for Plato is "The Good").[16]

Note here that a person only acts from the standpoint of the individual; a person is responsible for *only the instances* of universals, not the actual universals themselves. Human beings together have no collective role to play in the generation of universals. Universals just *are*—in much the same way as universals in logic and mathematics seem to just *be*.

Deities for Plato are also souls, and he includes among his deities a divine creator of sorts. Like other souls, this deity, the *demiurge*, does not create the eternal forms (i.e., "universals"). Instead, it fashions the sense-perceived universe out of the forms.[17]

In sum, for Plato, we are individual souls that instantiate properties, but neither individually nor collectively are we the source of the properties in themselves. Even properties like "beauty" and "virtue" that we would today think of as clearly human specific do not come from us. They reside in a realm separate from the everyday world of particulars.

Turning to Aristotle, the East-West difference in conceptions of reality is perhaps sharpest in his fundamental concept of "substance," or *ousia*.

Aristotle's concept of substance answers the question of what is the basis for all things. In Nondual Hinduism, the concept of Brahman answers this question by saying that there is only one basis in the universe, the aware "self" or "I" principle. It is this one Self that is responsible for all individual selves. For Aristotle, substance (or, more precisely, "primary" substance) answers the same question; however, it answers it very differently. Aristotle's substance identifies not what is *universal* to all individuals but rather what is *individual* to all individuals—what makes them what they are as individuals enduring over time and *different* from other things.[18]

Let me repeat this, since Aristotle's concept of substance has been centrally important in Western theology as well as much of Western philosophy. Aristotle's substance does not identify what is *universal* to all individuals but rather what is *individual* to all individuals. Each enduring particular has its *own* substance. Substance is that ingredient of nature responsible for particulars being particular and numerically just one thing. For example, the substance of Socrates is what made Socrates the particular person he was as opposed to someone or something else. Throughout

Socrates's life, something had to persist in order for him to have continued to be "Socrates." "Substance" names that which thus persists, and in Aristotle's philosophy, it largely replaces the concept of soul.

Like Plato, Aristotle has a concept of a supreme deity, but in his case, it is modeled on his concept of substance. While Plato's God is the soul that fashions the universe out of the preexisting forms, Aristotle's God is the substance of the universe as a whole, what makes the universe what it is as opposed to something else. The universe has a basis that cannot be taken away or changed without it ceasing to exist.[19]

Another important group of philosophers are the *Stoics* (third century B.C.E. to second century C.E.), who contributed, among other things, their concept of *logos*. Logos is a Greek term with a variety of meanings, such as "word," "speech," "principle," and "reason." The Stoics adopted the term to mean "reason" as the supreme cosmic principle responsible for the order of nature. For the Stoics, reason pervades and shapes the entire universe as a sort of divine will, which we glimpse through our capacity for reason.

The Stoics seem to have been the first important group to face the implications of first principles for the idea of human free will.[20] Because all things are governed by the logos, Stoics considered human life to be almost entirely predetermined. Free will exists, but only in one's personal values—one's choice to face one's fate with or without virtue. Thus while individuals cannot control the course of the universe, they can control their happiness according to how well they work in partnership with their fates. Free will and choice are not about getting what you want but about wanting what you get.[21]

Notice here the difference between Stoicism's concept of logos and Hinduism's concept of Brahman. Hindus speak of enlightenment as *liberating* us from karmic predestination. Joining our will to the will of the universe brings wisdom and yogic powers. But the Greek idea of the soul as individual in nature precludes such a concept. Collective agency and individual agency are fundamentally different. For the Stoics, joining our agency to the agency of the universe doesn't empower us so much as align our wills with the preexisting, separate rationale of the logos.

Saddled with a concept of soul very different from the one in Hinduism, notice also how much trouble these Greek philosophers have finding a basis for universals. Plato says they exist in another realm of their own. Aristotle says they exist in particular things. The Stoics claim they come from a supreme cosmic principle, the logos. Jews (for the most part) claim they are created by a supreme being more separate from the universe than a logos.[22] And the Skeptics, whom I touched on in prior chapters, claim that either universals don't exist or they are beyond what can be known.[23]

This conventional Greek concept of soul was not, however, the only option. The first century or so B.C.E. saw the beginning of Gnosticism (from Greek *gnosis*, "to know"), which taught concepts of soul more similar to the Hindu Ātman. Gnostic beliefs varied widely from group to group but were linked together by the idea borrowed from Greek philosophy that God's nature was related to human nature via human intelligence.[24] For Gnostics, God was, in various differing ways, human nature writ large as a first principle of the cosmos. Because of this affinity between God and humans, they believed that one could achieve direct knowledge (*gnosis*) of God and of one's kinship with God through the intellect.

Then later, in the third century C.E., the enormously influential philosopher Plotinus (204/5–270 C.E.) taught a mix of religion and philosophy that came to be called "Neoplatonism." Plotinus transformed Plato's idea of the Form of all forms into a supreme being called simply the *One*. Very much like Brahman in Nondual Hinduism, the One is the primal pattern, the archetype, for *everything* in the cosmos, including ourselves. From the One emerges the *nous* (intelligence) and then souls and matter.[25]

Thus at the center of Neoplatonism, as well as much if not all of Gnosticism, was human intelligence transformed into a basic principle of the universe. This was not the dualist philosophy of Abrahamic orthodoxy but rather a third-person strategy akin to Nondual Hinduism.

Abrahamic monotheism reacted to these third-person accounts of reality in a variety of ways. Judaism, Christianity, and Islam all had and continue to have mystical traditions that embrace nondual, third-person strategies to one degree or another.[26] Judaism developed a tradition and literature around these ideas, the *Kabbalah*, that came to influence Christians as well. Orthodox Christianity was probably the most strident in rejecting these ideas. It institutionalized the concept of *creatio ex nihilo* (also found in Islamic philosophy), which postulated God creating the universe out of nothing.[27] Still, even though this doctrine placed God's agency in a realm entirely apart from human agency as well as all of creation, it didn't stop certain well-known Christians from discussing the possibility of a mystical union with God.[28] Islam, for its part, combined the dualist monotheism of the Qur'an with Aristotelian philosophy, Neoplatonism, and Gnosticism to produce, over the centuries, a wide variety of beliefs ranging from the relatively strict Aristotelianism of Ibn Rushd (1126–1198 C.E.) to the Neoplatonic-leaning mysticism of Ibn-al'Arabī (1165–1240 C.E.), with most Muslims, not surprisingly, preferring a more orthodox scripture-based and faith-based theology.[29]

In general, the main orthodox strands of Abrahamic monotheism opposed nondualist strategies. God might have all the traits of an archetypal aware self like Hinduism's Brahman, but God was not Hinduism's Brahman. God and human agency were separate.

Human Nature and Life Goals

I began this chapter with three Abrahamic prayers that illustrate the dualism of interest to me here. All three prayers imply two separate kinds of existence or being: God's manner of being (which is typically beyond what we can know) and our own human manner of being.

The principal theme of these central, widely used prayers is not, however, the separation between God and humans but rather the *bridging* of that separation. While God and humans are thought of as two different kinds of agency, we are asked by these religions to *overcome this separation* as far as we are able. Per these prayers, we are to love God with all our heart, soul, and might; we are to affirm the coming of God's kingdom on earth; and we are to worship and seek help in only God.

In certain respects, this may seem similar to Hinduism's concept of moksha, whereby we overcome our everyday sense of separation from Brahman. But the concept of self is different in these Abrahamic cases and so, too, is the characterization of the goal. Our mortal predicament in Abrahamic monotheism is not that we are one with God and don't realize it. Instead, we and God are fundamentally different entities. We humans are as we discover ourselves in our everyday practical affairs. We are individual agents only; whatever the collective agency may be for the public nature of the universe, it's not us.

And yet we are asked by these monotheistic religions to overcome our separation from God. How is this to be done when we are thought to be of entirely different natures?

The short answer is that we cannot normally overcome this separation—at least not entirely—as long as we are part of a reality conceived of in this way. Generally speaking, Abrahamic orthodoxy depicts us as progressing toward a final goal that is not to be achieved within reality as we currently know it. Our final rest in heaven or paradise is thought of as coming *after* this mortal life. The final judgment, the rapture, the coming or return of the messiah or the Twelfth Imam, the restoration of God's kingdom, Zion, the promised land (the list goes on) are thought of as transforming life as we know it now; *reality as it currently is for us is no more.* Note the contrast to most Buddhist and Hindu beliefs in this regard. In these Asian traditions, we are repeatedly reborn into *this life as it is* until we get it right—until we are enlightened. It is not the world we live in—reality—that is transformed; it is ourselves and our minds.

Of course, there are exceptions on both sides; each of the world's major religions embraces many philosophical views. Pure Land Buddhism aims at the heavenly way station of the Western Paradise as opposed to an enlightenment in or return to this reality. Judaism tends to deemphasize the afterlife in favor of communally oriented, this-worldly goals in keeping with a relationship with God conceived of as less an individual matter than a collective covenant. And as I've mentioned, all the Abrahamic religions have mystical traditions that seek, in various ways, to collapse the self-God divide (often while simultaneously trying to keep it[30]).

Regardless of diversity, however, the root philosophical problem shaping all these religious views—East as well as West—seems to me to be the separation in our everyday lives between the standpoint of the individual (ourselves) and the standpoint of the collective (God, Brahman, mind, collections of minds). We live in a public world that is independent of us individually, yet not independent of us collectively. How we conceptualize and face this situation decides not just how we characterize the ultimate nature of reality but how we characterize human nature and life goals as well.

Summary with Strengths and Weaknesses of This Strategy

The distinction between individuals and collections of individuals has been central to my discussion of accounts of reality. If we trace our public world to the here-and-now awareness of individuals, we get first-person accounts. If we trace our public world to what is in itself true for all, we get third-person accounts. If we trace our public world to both these standpoints equally, we get dualist accounts.

Dualist accounts as I've characterized them here are anchored in reality as we find it in our everyday lives. In these accounts, human beings are also as we find them in our day-to-day lives. They have individual agency only; collective agency is thought to lie elsewhere in whatever aspect or aspects of a people's collective life seem best to explain the properties and orderliness of nature.

Polytheism, for instance, illustrates a straightforward, prephilosophy version by tracing collective agency to multiple deities identified with various important elements of collective life. Abrahamic monotheism, which emerged at roughly the same time as Western philosophy, replaces the multiple agents of polytheist deities with a single, universal authority more in keeping with that era's new understanding of the public world.

The concept of ourselves as having individual agency only is supported by much of Greek philosophy. Plato conceives of the soul not as the source of universals but as what fashions universals into particulars. Aristotle's substance is not what makes us the same as others but what makes us individuals. Stoicism's logos may predestine the course of our lives, but we are still individuals in our efforts to accept our fates. Exceptions to this Greek concept of self are found in Gnosticism, Neoplatonism, and the various Abrahamic mystical traditions, which teach that human nature is in some sense an expression of God's nature.

In a certain respect, Abrahamic religions share Nondual Hinduism's life goal of overcoming the separation between ourselves and God. However, their different concepts of self come with different characterizations of this life goal. Abrahamic orthodoxy (in contrast to its mystical traditions) considers our difference from God to be fundamental to the way reality is for us (at least in the current age)—something that we can only partially, but never fully, bridge through spiritual practice. Abrahamic mystical traditions, on the other hand, influenced as they were by Gnostic and Neoplatonic concepts of self, typically seek a more complete union with God through transcendence of the self as we know it in everyday life.

One of the great appeals of this dualist strategy is that it takes reality as we find it in our day-to-day lives. Our everyday experience of being enduring beings living in a world with other, separate-from-us enduring beings is not an illusion. With ourselves as individual agents only, no questions arise as to how we could also be the origin of the orderliness we find in the cosmos as a whole. This strategy also provides a ready answer to our enduring inability to settle the debates between the other two strategies. Everyday reality appears to involve both first-person and third-person views; to base an account of reality on just one or the other seems arbitrary and hard to justify.

In spite of such strengths, a great many philosophers find dualist accounts to be the least satisfactory of the three. Like the other two strategies, this strategy seems only partially right. It certainly may be true that an account of reality should accommodate both first-person and third-person views in a way that does justice to both. But surely our account should *also* explain how the two views relate to each other. It should provide some conceptual framework for overcoming the philosophical contradictions generated by the two views and, in the context of monotheist religion, for explaining why our spiritual practices should seek to bridge the gap between ourselves and God.

It's here at the interface of first-person and third-person views that we seem invariably to run into philosophical problems. If one of the two views isn't primary, then we seem faced with trying to explain how we can both have and not have free will; or how God might be our very life and breath yet also be an agent of the universe apart from us; or, since Descartes, how our interior mind might relate to exterior matter. With issues like these, nondual accounts usually seem the more palatable. They at least propose ways to resolve the dualism responsible for such questions.

This chapter completes my illustrations of three contrasting accounts of reality. Next, I want to consider how all three strategies might be justified—how they might be different solutions to a shared problem, different approaches to a particular mystery lying at the heart of our relationship to the cosmos.

I'll begin by reexamining the nature of awareness. Unlike the chapters on Hinduism and Buddhism, I haven't talked about awareness or consciousness at all in this chapter; rather, I've spoken about individual and collective agency. What, then, might be the relationship between awareness and these two kinds of agency—especially the collective agency that humans so often associate with deities?

Reality as Fugue

Introduction to Part Three

This book seeks to clarify a philosophical puzzle that appears to me to be responsible for certain long-standing debates over the nature of reality—a puzzle concerning the relationship between first-person and third-person views.

Central to my analysis has been the distinction between individuals and collections of individuals. I've described the first-person view as what an individual experiences and the third-person view as what remains the same for collections of individuals or matter. I've also discussed how this distinction between individuals and collections of individuals divides agency into two kinds. Individual agency points to the origin of what exists insofar as it originates with individuals (e.g., individual living beings, particles of matter) and their cause-and-effect interactions; collective agency points to the origin of what exists insofar as it originates with collections of such individuals.

While the difference between individuals and collections of individuals may seem fairly straightforward, our pictures of reality differ depending on whether we begin from the standpoint of the individual, the standpoint of the collective, or both. The contradictions among these pictures have persisted despite many efforts to resolve them.

Up to now, I've discussed this problem principally in terms of three contrasting accounts of reality. The remainder of the book considers these accounts as themes in a single composition. It examines more closely the puzzle itself and what the three accounts taken together might add to our understanding.

CHAPTER NINE

Awareness's Two Roles

Who Are We Really?

Just before I began this chapter, my brother Bruce and I had been talking about how his idea of "self" was different from the way his therapist used the word. As a consequence of his schizophrenia, Bruce had for a long time felt himself pitted against enemy voices that, as he now agreed, came from his mind. What he did *not* agree with his therapist about, however, was that these voices came from *himself*. Per his experience, these voices arose without his conscious intention, and many were enemies that he wanted to be rid of. It didn't make sense to him that what produced these voices and vexed him like this could be himself. For him, the self was something other than his mind and whatever unconscious mental faculty created his voices.

When I was younger, I became interested in what are often called out-of-body experiences, and I learned to produce them. An out-of-body experience is like a vivid dream, in that one has the sensation of interacting with and moving about in a world that seems extraordinarily real, although one's physical body remains insensate and dormant. Unlike a dream, however, one is awake and conscious both of what one is doing and of being disengaged from one's normal bodily perspective. I would intentionally leave my body, so to speak, so that I could see it from the ceiling, for example, then travel elsewhere and later return (although in one anxious session, the body to which I repeatedly tried to return kept turning out to be another out-of-body experience).[1]

Out-of-body experiences are so vividly real that many people believe that a soul or "astral" body literally leaves the physical body to travel elsewhere. On this matter, I agree with those who say that our minds produce the world of these incorporeal travels.[2] Nevertheless, like Bruce and his voices, I certainly don't experience myself to be the one producing that world. I may induce the mental state and have some control over where I go and what I do, but I don't experience the world in which I find myself as coming from me or my awareness. Whatever the faculty of mind may be that constructs these out-of-body circumstances, that faculty operates unconsciously and is something apart from the first-person conscious self who experiences and interacts with these circumstances.

I've also discussed in prior chapters how the everyday public world of our first-person experience seems to depend on our minds collectively while being independent of our minds individually. On the one hand, we have abundant evidence from a variety of sources (for a review of this evidence, see text box) that what we perceive around us as a shared world is a presentation of our minds; it's something generated within our first-person views as that which might exist for others as well as ourselves. On the other hand, we don't typically experience *ourselves* as the source of this world that our minds show us. Our everyday world is presented as something whose existence doesn't depend on us, as something that is true for collections of individuals or matter irrespective of our own personal awareness of it. In our everyday experience, again, we find a difference between the unconscious faculties of mind that present us with this world and the conscious self to which this world is presented.

So then, given these examples, who are we *really?* Are we each our mind with its unconscious processes, such as that which presented Bruce with his voices, presents me with the world of my out-of-body experiences, and presents all of us, it seems, with the everyday world of our first-person perceptual experience? Or are we each a first-person conscious self that is subject to what the mind thus presents?

I've already discussed different answers to this question. In my Hindu and Buddhist illustrations, our everyday idea of self is, in the final analysis, an error. The public world we're conscious of around us is not something to be understood apart from the faculties of awareness that present it to us. In our true nature, we are what generates our experience of self and world—be it an archetypal superconscious mind "dreaming" (so to speak) our universe or be it interdependent, here-and-now occasions of awareness out of which our universe and sense of self arise. On the other hand, as opposed to these Asian alternatives, Abrahamic orthodoxy as well as most of us in our everyday lives consider as paramount the *difference* we experience between ourselves and the world we live in. Who we really are is not what creates the world that is presented to us—God or mind or brain or whatever else that source may be— but rather the conscious self or soul who is subject to the public world created by that source.

This debate also concerns agency. For Hindus and Buddhists, the knower and the object of knowledge are linked together, and the same agency that produces

THE MIND DEPENDENCE OF OUR PERCEIVED WORLD

The Enlightenment and the centuries after have given us varied, compelling reasons to conclude that much if not all of the world we perceive around us is a creation of our minds. What we're aware of is not something apart from the way we're aware of it. This evidence is scattered throughout the book; let me summarize it here.

1. **The physiology of perception.** If, for example, we saw only what our eyes showed us, we'd see just patterns of light, if even that. This difference between the data presented to our sense organs and what is present to us in first-person, perceptual awareness implies that our cognitive faculties create the public world of our perceptual experience, and that what we perceive is either a mental representation, a mental invention, or some mixture of the two.

2. **Illusions and hallucinations.** Perceptual experience by itself is prone to mistakes and deception; what we perceive can even be a dream or hallucination. We determine what in fact exists by turning to reason and logic, which are faculties of our mind.[3]

3. **The historical difficulty of explaining universals in terms of sense experience.** There would be no world as we know it without universals, but universals don't seem to be something we can know through our direct sense experience of the world. Plato observed that we apprehend universals not with our sense organs but rather with our intellects, and even empiricists like John Locke and David Hume didn't claim that we directly perceived mind-independent universals. If this is the case, then we seem left with all universals—if not all the particulars that they describe—originating largely if not entirely within our minds.

4. **Anthropology and sociology.** Comparative anthropological studies show not only the dependence of worldviews on cultural

context but the dependence of our understanding of *anything at all* on language and other cultural inheritances.[4] While there is controversy over the universality of cultural structures,[5] no one disagrees that our understanding of the world is shaped in some sense by a mosaic of overlapping and interconnected collective memberships—memberships that range from family to friends and professional associations, to caste or class, to ethnicity and country, to religious affiliations, to linguistic communities, to species, and so on. We are, in other words, creatures of our collective memberships, with these collections varying enormously in scope, composition, and influence.

individual experience and behavior also produces our collectively shared cosmos. Individual and collective agencies are both expressions of the way our awareness generates a public world. Abrahamic orthodoxy, on the other hand, stays consistent with our everyday experience and characterizes us as individual selves or souls and agents in our own right. Our individual agency is distinct from the agency of God, the collective agent who sets the rules for the public world of our everyday experience.

At the heart of these alternatives lies the puzzle of interest to me. It's a puzzle about human nature and reality, but principally about the nature of awareness.

Awareness's Two Roles

Notice in my examples of the prior section the two pieces of this awareness puzzle. On the one hand is our everyday conscious self; on the other are the unconscious mental processes that present us with a seemingly separate-from-us public world.[6]

Consider the different roles played by awareness in each piece of the puzzle. First, as we experience in our conscious awareness, it plays an individual or "disjunctive" role. Awareness provides a here-and-now, first-person arena for experiencing and interacting. It is "disjunctive" in that such a first-person view means being here and not there, being in the present time as opposed to another time, being just one person and not many.[7]

Second, as evident in the way our unconscious mental processes present us with a seemingly separate-from-us public world, awareness plays a collective or "conjunctive" role that's largely unconscious. While the disjunctive role provides the first-person occasion of awareness, this conjunctive role provides the content—the properties (i.e., publicities) that associate what we're aware of now with what we and/or others are aware of at other times and places.[8] This role is "conjunctive" in that it accomplishes its task by grouping together different here-and-now, first-person views according to what they hold in common. These groupings are distributed not just in space, as, for example, all English-speaking people. As I observed in chapter 7, they're also distributed in time—for example, the collection of here-and-now moments of experience that make up the life of a single aware being.[9]

Notice the tension between these disjunctive and conjunctive roles; they relate first-person views to each other in opposing ways. Awareness's disjunctive role accomplishes its task by being a here-and-now, first-person view that is *different* from others. It's perspectival; it's this place and time, not some other place and time. Contrariwise, awareness's conjunctive role accomplishes its task by being a here-and-now, first-person view that is *not* different from others. In order to present properties that are recognizably the same from one occasion of awareness to another, first-person awareness must obey certain shared rules. The process of awareness must remain sufficiently alike from one instance to another to produce the same properties under similar circumstances.

Working together, the two roles generate the everyday reality of our first-person view; they bring into existence the public world that we experience around us. The disjunctive role generates presence—our perspectival, here-and-now occasions of awareness, irrespective of any content (other than presence itself[10]). The conjunctive role adds the content, the publicities. It ascribes properties according to what might be the case for more than just one's own here-and-now awareness.

In a certain respect, this analysis of awareness fits well with what we've come to usually mean by first-person view since Descartes. As I mentioned earlier, there is abundant evidence from a variety of sources that the world we're conscious of around us is generated by our own mental faculties. Whatever the clock on my desk may be in itself, what I personally see is the clock presented by my own mind, the clock as I myself personally experience it, the clock that is there only when I'm conscious of it.

In another respect, however, my analysis differs from our conventional one and modifies certain of its historical difficulties. I've discussed how this post-Descartes understanding of first-person experience has been at odds with our everyday experience of directly perceiving public objects *in themselves*. When I look at the clock before me, I don't think of myself as seeing a clock that's there only when I'm conscious of it, a clock that is generated by my own mind. I'm presented instead with what seems to be a third-person clock, one that is independent of my mental faculties and there for others also to perceive. This discrepancy between our post-Descartes understanding of first-person awareness and first-person awareness as we individually experience it is, if you recall, one of the most vexing philosophical legacies of the Enlightenment.

If, however, we grant that properties originate with collections of occasions of awareness or matter (they're publicities rather than merely universals[11]) and if, further, we divide the activities of awareness into the two roles proposed here, then it seems to me much easier to understand how our faculties of awareness might in fact present us with a genuinely public world. Our individual acts of perceiving public properties are then *also collective acts* of generating public properties. Our process of awareness doesn't generate simply an individual experience of reality; it also participates in generating a collective reality for us to experience.[12]

With awareness conceived of like this, the reason we are personally presented with a genuinely public world is that the process by which awareness generates this world doesn't belong to just you, or me, or one moment alone. We directly perceive what others can directly perceive because our first-person views follow certain collective rules of thinking, perceiving, and interacting given by the humanness of our bodies and brains, the biology of our cells, and the physics of our atoms and molecules (for a start). It is by awareness following such rules that we're presented with a world shared with others whose awareness follows the same rules. I directly perceive a clock that other people directly perceive because I'm the local human representative of a collective way of being aware that perceives certain hunks of matter as "clocks."

As should already be clear, this account of awareness is based on properties having many sources. Some properties come from matter and living things besides humans (e.g., the properties of atoms or trees insofar as they're independent of the human mind). Some originate entirely or almost entirely with humans (e.g., the property of being Santa Claus, the monetary value of a Degas painting). Some originate only partially with humans (e.g., the property of being a teacup or clock as an object made out of matter by people). Some originate with human beings but refer to properties that originate elsewhere (e.g., the atomic weights of atoms as a numerical property invented by humans to signify the relative mass we observe in atoms). Some properties originate with and describe individual aware beings themselves (e.g., the distinctive physical features and behavioral habits a particular person develops over time). The next chapter examines some of the basic ways occasions of awareness group together to generate public properties.

I also want to make clear that my division of awareness into two roles doesn't intend to solve the Cartesian gap problem so much as relocate it as a problem of awareness. The puzzle of how awareness can be both individual and collective at the same time remains, and is for me, the much more fundamental issue, a puzzle from which arises not only the Cartesian gap but conflicting accounts of reality and of who or what we ultimately are as aware beings. It also seems to me to be the source of a number of other philosophical problems, some of which I've already looked at (e.g., Goodman's "grue" puzzle of inductive logic in the chapter 7 text box) and some of which I'll look at later.

Among the implications of this puzzle of awareness are also the differences among religions that I've focused on here. To one degree or another, we live in a public world that is independent of our individual conscious awareness yet dependent on our mostly unconscious collective awareness. How we conceptualize this situation determines our understandings of self, world, and deity, as well as our life goals and spiritual practices.[13]

Sitting in a Café

Let me review and expand on these activities of awareness with the following illustration. First I'll look at awareness's conjunctive role, then awareness's disjunctive role.

I'm sitting in a café by myself, eating breakfast and looking around at the people, furnishings, and decorations. Customers arrive, talk, eat, and leave, while the staff fills orders, brings food, and clears tables.

As I look around, I reflect on the relationship between my mind and what I perceive.

I notice how the things that I perceive present themselves in my consciousness as "public," as characterized by properties that others can also perceive. I have no doubt that other customers besides myself can read the chalkboard menu hanging on the wall and can order their meals from the same waiter that I have. Should I drop and break a plate, surely it would be broken not just for me but for others in the café too.

Of course, the curious mouse or spider wouldn't likely perceive the mess on the floor as a "broken plate," but such creatures are not major factors in the publicity of most of what I perceive. For me, the relevant aware beings at the moment are people—human beings with faculties of awareness similar enough to my own for them to perceive essentially the same things that I do.

Physical properties—properties that originate with matter—are also important to the public nature of what I perceive, but they seem invariably presented to me in terms of how they relate to human capabilities and interests. I don't know the room's exact spatial properties, and in any case, they're not especially important compared to other, more human-oriented spatial features—the height of the tables as comfortable for eating or the proportions of the room in terms of function and aesthetics. Likewise, the material compositions of the things I see are expressed not as atoms and molecules but as the tastes of food, or the colors of the walls, or the odors from the kitchen, or how the coffee maker works to make coffee. This does not mean that physical properties in themselves are not important; rather, it means that they are not presented to me in themselves. Even when I use a tape measure to give numbers to spatial distances, the tape measure is itself a measuring device suitable for our human way of being aware.

So far, the properties I've distinguished describe the room around me as I presume it appears to not just me but other people as well.

What, then, about properties that seem to apply to me alone and the way I myself perceive these things? What about the food as it tastes to me personally or the aesthetic qualities of the decor as I, myself, experience them? Perhaps I taste a distinct but unfamiliar flavor that I couldn't describe to someone else but that I might—or might not—remember later. The properties in this case are personal; they're private and apply to just myself, not to anyone else.

As I think about private properties like these, I note that they, too, have a collective context. They, too, are what I've called "publicities," but their context doesn't involve other people besides myself, just me personally—myself as a temporally enduring being.

In other words, the collective context in this case is in *time*. The properties of my here-and-now experience of this café rest on my past experiences of other cafés and, beyond that, on the many other here-and-now experiences that make up my life. Later, when I leave here, these subjective elements of my experiences will be part of my lifetime of experiences. An especially pleasant smell coming from the food on my table today will become part of my memory tomorrow. The vividness of the experience may fade; my memory of it may disappear entirely. But to the extent and in whatever way it endures, it will remain a property of what I once experienced at this moment. The smell will become one experience located within the collection of all my experiences, part of my stock of personal properties.

Apart from properties like these, all of which come with collective contexts (some private, some not), what else might I be aware of? I've been focusing on awareness in its conjunctive role. What about awareness's disjunctive role and presence?

I look around again at the people, tables, chalkboard menu, and the rest. All these things that I see are present to me in my conscious awareness. I close my eyes and listen to the sounds around me and what they identify: forks on plates, people talking, cars going by outside. All these are present to me. The couple at the next table are talking about an upcoming wedding. What I hear of their conversation is present to me, and through their conversation, the two of them are also present to me as aware beings with their own cares and interests.

All these things are present to me, as is everything I'm conscious of. They all have the quality of "presence in my awareness," and nothing that I'm conscious of is without this quality.

Yet as I reflect on this property, I'm also aware that this "presence" is not a property like other properties.[14] It doesn't differentiate one thing I'm conscious of from something else that I'm conscious of; *everything* I'm conscious of has presence in my awareness. Presence is more the arena in which properties make their appearance, more what transforms latent properties into ones that are expressed. Unlike the usual properties presented in my consciousness, presence is more precisely a property *of* these properties.

While presence may not differentiate one thing I'm conscious of from another, it does come into play in my judgments about what might be present to others or present to myself at other times and places. Out of all the possible sights and sounds in this room, I'm consciously aware of only some of them.[15] I can't hear everything that the couple next to me are saying; I don't see, or hear, or consciously attend to all that they or others see and hear. What is present to others, the properties they respond to, is not the same as what is present to me.

Indeed, to think of all the possible people and other creatures that might visit this room is to think of this room as characterized by countless properties that I'm not currently aware of in any way at all, that are what I've called "latent," or merely capabilities for presence. And though these properties are hidden from me, they nevertheless play a critical role in the behavior of other aware beings and in how circumstances around me unfold. It's not the properties presented to *me* that determine what other aware beings do but rather the properties presented to *them*. And if I'm to predict the behavior of other aware beings, it's these properties—the ones that they each in their own way are aware of—that I need to understand.

These critical yet hidden capabilities for presence add depth and mystery to what I perceive, a complex texture of unknown possibilities that I also think of as part of my everyday world. What is currently present to me—this room I'm sitting in, for example—always seems to have a context that's larger than what is currently present to me. It always seems to arise against a background that encompasses not just my here and now but the here and nows of my past and anticipated future, as well as the here and nows of the pasts and anticipated futures of others. This café room is not *just* the room I'm conscious of at this particular moment. It is that, of course, but it's *also* all my many experiences of cafés rolled into this moment and, beyond that, cloaked in this moment's latent properties; it's all that this room might be to *any* aware being.

The Consciousness Problem

Next, let's consider awareness in the context of current debates on the subject of consciousness. Let's begin with the concept itself and the distinction we make between conscious awareness and awareness that is not conscious.

Recall from chapter 5 how the word *consciousness* distinguishes a certain attentive or focused kind of awareness that we have when we're awake. When we are not awake, we speak of ourselves as "not conscious," even though we are aware enough of our surroundings to wake up if we hear a loud noise. When we are awake, we're conscious of only a small part of everything we're aware of. Without our conscious attention, we monitor background sounds, sights, and smells for anything that might be of conscious interest. If you're talking with someone at a noisy cocktail party, you typically will not be conscious of other conversations; however, your conscious focus will quickly shift if, for example, you hear your name mentioned.

Consider how this distinction between conscious and unconscious awareness fits with the two roles of awareness discussed earlier. When we speak of conscious awareness, we usually mean the combined activity of both roles (however, see the "'Intransitive' versus 'Transitive' Consciousness" text box). The disjunctive role generates the individual, here-and-now arena of awareness that is conscious awareness's distinctive "attention" or "focus"; the conjunctive role generates that of which we're conscious—the thoughts, feelings, perceptual objects, properties, and so on presented in first-person conscious awareness. On the other hand, when we speak of unconscious or subconscious awareness, we imply only the conjunctive role—awareness considered apart from the attention or focus provided by the disjunctive role's presenting process. The conjunctive role comprises the cognitive processes that coordinate different occasions of awareness and give us awareness's conscious content, and these processes are entirely unconscious except insofar as they receive the disjunctive role's attention. They operate entirely in the background except for that portion presented within awareness's conscious focus. When we look at a book or a person, we don't see *how* we see it or everything that our awareness has available to be seen; we see just the book or person presented in our first-person view.

Apart from giving us another way to understand the distinction between conscious and unconscious awareness, what else might this chapter's analysis of awareness offer?

"INTRANSITIVE" VERSUS "TRANSITIVE" CONSCIOUSNESS

Efforts to understand consciousness have distinguished a number of ways in which the word is used. Most of these usages are associated with particular theories of consciousness (e.g., consciousness means "being sentient," or "being sentient and wakeful," or "being aware of being aware"). In the context of awareness having disjunctive and conjunctive roles, David Rosenthal's distinction between "intransitive" and "transitive" consciousness is particularly worth considering.[16]

"Intransitive" consciousness is consciousness considered irrespective of its content as, for example, when we speak of someone who has just awakened as being "conscious" irrespective of what the person might be aware of. "Transitive" consciousness, on the other hand, is consciousness *of* something, which is how I describe it in the text. Consciousness in this sense includes whatever is present in our awareness—the feelings, thoughts, or perceptions *of which* we're conscious. In the context of awareness's two roles, intransitive consciousness highlights awareness's disjunctive role, which, as I've pointed out, is the role we experience as our own conscious awareness—that aspect of awareness we associate with ourselves and our first-person view. Transitive consciousness, on the other hand, is both roles together. It includes awareness's conjunctive role, whose process of generating our perceptual experience is largely unconscious. I may be conscious of various objects around me, but these objects seem separate from me, and I'm not conscious of how they become presented to me as part of the third-person world.

Consciousness in its intransitive sense—human awareness in just its disjunctive role—is especially important in religion. For example, the Sāṃkhya and Yoga schools of Hinduism propose a dualist philosophy with matter (*prakṛti*) distinct from individuated points of consciousness (*puruṣa*).[17] Also, meditative practices often aim toward a state of consciousness that's spoken of as being void of content (except perhaps presence itself). A term often used in this context is "pure" consciousness—consciousness that is still and empty while not in a state of sleep.[18]

The most troublesome problem of consciousness for philosophers and scientists these days has been that there is something about conscious awareness that doesn't seem explained by the brain's neurological structures and behavior.[19] This is not to say that research into the brain can't tell us about consciousness; rather, what we learn in this way seems to miss much of what consciousness is for us. Thus an fMRI or PET scan of a person's brain may give indications of conscious activity, but consciousness as we know it ourselves doesn't seem to be something that we can directly observe like this. In this respect it's like pleasure, pain, joy, or other subjective experiences. An fMRI scan may show us that the subject is experiencing pleasure, but it does so by showing us the relative degree of metabolic activity in different parts of the brain, not by showing us pleasure itself. What we observe in someone else's brain as "pleasure" is not pleasure as we know it in our own experience. Likewise, what we observe as "consciousness" in someone else's brain is not consciousness as we know it ourselves.

No doubt at least part of this problem of consciousness stems from the gap between interior mind and exterior matter that we've inherited from Descartes and the Enlightenment. On the one hand is what we know of subjective states and the nature of awareness when we examine them from the "inside," so to speak—what we know "in our minds." On the other hand is what we observe of them from the "outside," what we empirically discover by examining human physiology and behavior. These two perspectives give us two sorts of seemingly disconnected data with no obvious way to bridge the gap between them.[20]

David Chalmers, like many others, puts this difficulty in terms of first-person versus third-person views: "The fundamental issue concerns how to integrate two sorts of data about the mind. We have 'third-person data' about the brain and we have 'first-person data' about subjective experiences. Both are equally real, and both need to be explained. The task of a science of consciousness is to integrate them into a single framework."[21]

This chapter has proposed another perspective on the Cartesian gap. It has reframed this first-person versus third-person divide such that differences in data, such as the one found in this problem of consciousness, are consequences of properties having *different sources*. What Chalmers calls "first-person data"—that is, subjective experience—concerns properties presented specifically by and from within our own awareness, properties the source of which is ourselves and our shared way of being aware.[22] What he calls third-person data concerns the neurology of our brains—properties whose source is (principally) biological cells.

From this perspective, Chalmers's first-person data are just another kind of third-person data, and the problem of consciousness is principally a question about the relationship between two different sources of third-person data. By what mechanism

REALITY'S FUGUE

might collective agents of one kind (cells) group together to form larger-scale systems (in this case, individual human beings) that are capable of operating as collective agents within yet another even larger-scale system (human beings collectively)?

Reframing these two sorts of data in terms of their different collective origins is just the beginning of an answer, however. The next two chapters examine some of the basic groups involved in collective agency as well as how they relate to each other, to awareness, to everyday reality, to mathematics, and to the properties of matter.

Awareness also continues to have a mysterious side in this terminology, but for reasons that are perhaps clearer than before. The Cartesian gap reemerges in this vocabulary as a divide between the two roles of awareness and a puzzle of how awareness can be both individual and collective at the same time—a puzzle that will occupy much of the rest of this book. Then, too, recall how the disjunctive role expresses itself as presence, which is at least partially ineffable.[23] I will also have more to say about the ineffability of presence.

Summary

Various abnormal and altered states of mind like schizophrenia and out-of-body experiences illustrate how our everyday awareness is able to play two seemingly opposing roles at the same time. On the one hand, awareness plays an individual, or "disjunctive," role; it provides a first-person occasion of awareness. On the other hand, it plays a collective, or "conjunctive," role in which shared awareness processes generate a seemingly separate-from-us public world. It is because public properties are collectively produced that we experience ourselves directly perceiving a public world. I directly perceive a car or tree that other people directly perceive because I share with other people a collective way of being aware that perceives certain hunks of matter as "cars" or "trees."

The product of the disjunctive role of awareness is presence, which is less a property than the arena in which properties arise. To the extent that we might use it as a property, it's not so much a property of what one is aware of as a property of the properties of what one is aware of.

The products of our conjunctive role of awareness are "publicities," the third-person universals and particulars that originate from and are maintained by our ways of being aware. Publicities have innumerable sources besides just our own human ways of being aware; they also originate with other living creatures and matter. They can also originate with awareness as a progression over time, such as the sequence of experiences that make up the lives of individual people.

What we call "conscious" awareness is typically the combined activity of both roles as we find them in ourselves.

Dividing awareness into two roles reframes the problem that philosophers and scientists have had reconciling our first-person understanding of subjective experience with our third-person understanding of the brain. The difference between first-person and third-person data becomes a consequence of properties having different sources, and "first-person data" become just another kind of third-person data. From this perspective, the problem of consciousness has two parts. First is the question of how collective agents of one kind (beings with consciousness) might relate to and even depend on collective agents of another kind (biological cells) where the members of one collection are separated from those of the other by scale. Second is the question of how we are to think or talk about awareness's two roles when their functions seem mutually contradictory and the disjunctive role expresses itself as an at least partially ineffable presence.

CHAPTER TEN

Artifacts of Awareness

A Definition for Awareness

Early in this book, I argued that certain long-standing contradictions among accounts of reality would not likely be resolved without new concepts. To that end, I developed a definition for everyday reality (as distinct from ultimate reality) that better accommodates the philosophical accounts examined here than the more traditional definition that links reality to mind independence. What is real in this everyday sense is that which combines publicity with the capability of presence, with "publicity" and "presence" being themselves formal terms developed here. Defined this way, the concept also seems to help us better understand a philosophical puzzle that we come up against when we probe into the origin of the third-person view, a puzzle that divides accounts of reality into contradicting options.

Next, I want to introduce a formal working definition for awareness in order to help clarify what I mean by the term. This definition is based on the previous chapter's distinction between two roles of awareness coupled with the above concept of everyday reality. The definition seeks to characterize awareness in terms of what it does versus what it is; it doesn't address how awareness accomplishes its tasks. Nor does the definition necessarily capture all that awareness does—just those features I want to focus on here.

I'll give the definition first and then explain it. The shorthand version of the definition uses "presencing" rather than "presenting" in order to emphasize the particular sense of presence indicated in the paragraph following the definition and discussed in preceding chapters.

AWARENESS: The presencing of publicity. It (1) establishes presence (its disjunctive role), (2) cooriginates/comaintains publicities (its conjunctive role), and (3) joins publicity with presence to produce everyday realities (both roles together). All three elements are interrelated; anything that acts in one of these ways is aware and acts in all of these ways.

By "establishes presence," I mean that awareness in its disjunctive role differentiates itself as one particular here and now in causal interaction with other here and nows. As we experience in our own conscious awareness, it establishes a first-person[1] arena for the arising of properties; it expresses itself as a current, individual player in the universe.

By "cooriginates/comaintains publicities," I mean that awareness in its conjunctive role is an agent of publicity; it brings into existence and maintains first-person content that, as far as possible, remains the same over different occasions of awareness. Publicities can occur in time as well as in space; they can be private (e.g., a property we alone associate with an object) or shared by others (e.g., a property we as well as others associate with an object). Success in this activity requires a way to coordinate one here-and-now occasion of awareness with other such occasions. We are conscious of the conjunctive role principally through its products; in our everyday experience, we are not typically conscious of the mental processes responsible for what we're conscious of.[2]

By "joins publicity with presence to produce everyday realities," I mean that both roles of awareness act together to produce what is simultaneously public and present and therefore "real" in the everyday sense discussed earlier. Awareness is not simply an act that occurs in one moment isolated from other moments. It is simultaneously a *collective* act that produces properties that carry *collective* potentials for presence. As humans, our way of being aware collectively produces properties capable of presence that I've called "human specific." For other living beings, insofar as they possess awareness as I've defined it here, their ways of being aware collectively produce properties capable of presence that are specific to their kinds of awareness.

In regards to human beings, examples of mind-generated, collective realities are everywhere we look. Think of my previous examples: political borders, the value of money, the meanings of words, laws and legal contracts, clocks as ways to tell time, houses as places to live or social statements, trees as shade or lumber or firewood or decoration, animals as sources of food or pets, books as things to read—not to mention the more obvious examples of the shared artifacts of our culture (language, literature, philosophy, science, and religion, as well as our shared senses of beauty, ugliness, virtue, success, and failure). Indeed, *everything* that makes up the everyday reality of human beings that doesn't originate with other living things, matter, or the universe as a whole (in whatever form[s] that might take) is, as far as I can tell, generated by the human mind behaving collectively.

Finally, because the two roles of awareness are different elements of a process that requires both elements, to have one role is to have the other. Awareness, as I define it here, expresses itself as an amalgam of publicity and presence; it bridges the gap between one moment of awareness and many. I as an *individual* am aware of an object as a cat or a teapot because it looks like *other* cats or teapots; it fits a *collective* model *I* share with *others* concerning such things.

I want to draw your attention to the relationship between awareness and *existence* that I've proposed here. When I say that awareness produces everyday realities, I mean that *awareness literally creates the existence of things.* The generation of reality by awareness *is* the generation of existence.

Insofar as individual first-person views are concerned, this dependence of what exists on awareness should be uncontroversial and fairly obvious—it's what we typically mean by "first-person view" in our post-Descartes era. Whatever the neurological processes of my brain present in my conscious awareness—the clock on my desk, a thought, a dream—that's what exists in my first-person view. Stop these processes and the clock, thought, or dream no longer exists in my first-person view. Likewise for whatever is presented by ways of being aware other than our own. Although I might perceive an object on my desk as a pen, if a cat perceives it as something to play with, then that's what it is in its first-person view. That's what exists for this individual cat.

But the here-and-now properties presented within single first-person views are not the whole story of awareness's involvement in existence. As I've repeatedly illustrated, everyday reality depends on properties generated by awareness behaving *collectively* as opposed to individually. Take away a collective way of being aware (e.g., that of human beings, or bats, or individual aware beings as a sequence of experiential moments) and you take away the collective properties that accompany it. Introduce a new way of being aware into our universe and you introduce the collective properties that come with that variety of awareness. Certain properties, such as those of matter and logic, may seem to exist independently of awareness (a subject I'll take up in the next chapter), but all other properties seem to trace their existence to collections of first-person views and awareness's collective activities.

Note here that my tracing of existence to awareness doesn't mean that I've abandoned third-person philosophy. Reality continues to be what is so for all, irrespective of whatever individuals may be aware of. But "so for all" here doesn't refer to some theoretical "all observers" that is the same in every case. The picture I'm proposing is rather a mosaic of groups: some overlapping, some separate, some nested within each other. Each group generates its own reality and also contributes to the realities of other groups and to the gestalt of all groups together, with this gestalt, in turn, expressing itself through all its elements (though "gestalt" seems an inadequate word for whatever overall collective agency is at work here). It's this picture of reali*ties* (in the plural) that seems to me to promise a greater understanding of our universe, ourselves, and our religions, as well as how different philosophical strategies—monist as well as dualist—might in fact complement each other.

This mosaic of groups is my next subject.

A Mosaic of Groups: CORs, Classes, and CODs

This section examines collective agency: the dependent relationship of everyday realities (i.e., publicities with the capability of presence) on various collections of individuals.

I make my observations in a series of examples. In these examples, I refer to a collective origin of publicity as a *collectivity of reference*—a "COR" for short, with the plural "CORs" standing for collectivi*ties* of reference.

I call them collectivities of "reference" because we turn to them to understand the source of something that is public. As I've illustrated, the properties of the things around us point back to groups of origin; as part of their meaning, properties come with a "reference" to collections of living beings, or matter, or temporal collections of awareness occasions. I think of my desk as "useful for writing," but I don't think of that property as applicable to spiders or rabbits. It's a property that is given meaning by, or relates to, or "refers to," human interests. Thus if I wanted to understand desks, I'd examine human beings; if I wanted to understand anthills, I'd examine ants.

The concept of COR needs to be distinguished from the concept of *class*. A class is the collection of a publicity's instances (all desks, all anthills), while a COR gives an origin of that publicity (human beings, ants). The property of being a desk groups together all desks into a class; it also identifies human beings as a COR.

As I'll illustrate below, classes differ from CORs in several ways. One way is that a class can have multiple or an indefinite number of members (e.g., all wooden desks or desks in general), or it can have one member (e.g., the desk in my office), or it can have no members (e.g., the desk on the roof of this house I'm sitting in). A COR, on the other hand, must have more than one member if it is to be a source for what is the same from one member to another.[3] A single human being can be a COR, but only as an enduring being encompassing more than one occasion of awareness.

The following examples begin with everyday realities, such as a desk, where humans form at least one COR. I'll end by discussing realities without humans as a COR.

Material Human Artifacts like a Desk, Coffee Cup, or Clock. Consider the physical human artifacts that I've already talked about at some length: desks, clocks, coffee cups, cars, pens, and so on. These are objects with two primary CORs: human beings and matter. As mentioned above, their classes can have many, one, or no members depending on whether we are talking about a collection of things, a particular thing, or things with no instances.

For most of us, as long as we're talking about a class with instances, these artifacts fit with what we would usually call "real" in our typical everyday use of the word. Certainly, their physical properties seem independent of the human mind. And while human-specific properties like "deskness" or "clockness" may not be independent of our minds collectively, they do seem independent of our minds individually. In any case, both these material and human-specific properties are real in the sense stressed here of combining publicity with the capability for presence. They are shared aspects of our collective experience with the potential for expression in individual, here-and-now interactions.

Especially important is how the reality (as capabilities for public presence) of such artifacts depends on *both* CORs. Remove all human minds and the clock on my desk would participate in the world only as a hunk of matter, not as a clock. Remove the matter responsible for its physical properties and it would be only a concept, an imaginary clock, real perhaps as an idea but not as a physical object. Such artifacts exist not just because of matter but also because of what human awareness adds to matter. These added properties don't exist in matter by itself but in *our relationships* to matter.[4]

The realities of our everyday lives need not, however, have matter as a COR. Consider my next example.

The Rules of a Housing Community. While the human artifacts just discussed have both human beings and matter as CORs, many other artifacts have CORs made up of only human beings.

For example, consider the rules of the housing community in which my wife and I live. The houses here are individually owned condominiums, while the property around them belongs to the community as a whole and is cared for by a manager and staff. The community has various rules that govern, for example, what colors we paint the outside of our houses, what options we have for window and door styles, payment of dues, and how we choose our board of directors.

Observe how these community rules (as well as, more generally, laws, word meanings, collective values, the value of money, etc.) are not like clocks or desks. They interact with us not as physical objects but rather as concepts or ideas. When I buy paint for the trim of our house, I'm aware of the rule governing the color of the paint, but there's nothing physical about the way that rule expresses itself in my encounters with it. These community rules present themselves to me as things that are strictly mental in nature; they identify classes of instances that don't have matter as a primary COR, only humans. Although they're written down (and all members have a copy), the rules are not themselves these physical copies. They would still be rules to follow were they maintained by our memorizing them.

But if these rules are only ideas in our minds, are they in fact "realities" of our universe?

While these rules are not real in the sense of being mind independent, they're certainly real in some sense of the word for all of us living here. The rules hold for us and impact our activities regardless of whether we agree with or even know about them. If I painted my house the wrong color, our community manager and board of directors would pressure me to repaint it the right color regardless of whether I liked the color or had known about the rule. In the context of the terminology proposed here, these rules are "real" because they are both public and capable of presence. They are public in that all community members share them, and they're capable of presence in their direct effect on our here-and-now individual behavior.

Let's return a moment to the idea of humans as a COR for these rules. What about matter and the cells of our bodies as CORs? Without them there would be no human consciousness much less community rules.

I have no doubt that human consciousness depends on the behavior of brain cells. However, the realities under consideration are the rules of our housing community, not occasions of human awareness. While individual cells no doubt play roles in behind-the-scenes CORs, they do not form a *primary* COR for these rules. They do not directly participate in the CORs that present the realities of interest. Human artifacts such as laws, languages, money, political borders, philosophies, religions, and so on trace to the collective behavior of people as whole individuals. When people get together to talk about politics, religion, or housing community rules, it's not individual biological cells that discuss these matters but rather creatures made of trillions of cells. It's a *system* of many, many cells that composes an individual and a COR member, not a cell in itself. (In the language of sets, the COR for these community rules is not a set of cells but rather a set made of sets of cells wherein the subsets are the individual human beings.)[5]

These questions of what beings belong to which collection and what collections are primary versus secondary are central for understanding realities as public capabilities for presence. Recall from prior chapters how the perception of properties relies on sorting out who is aware of what: what properties originate with ourselves individually, what properties originate with various groups of people, or what properties originate with other creatures. This is also true across different scales. Human consciousness may depend on brain cells; however, the public properties of our world that are presented to us in our first-person awareness have little to do with brain cells acting individually and much more to do with them acting collectively in a way that's consistent from one human being to another.

The People Belonging to Our Housing Community. The members of CORs themselves differ from artifacts by participating in their own CORs; the class is itself a primary COR.

Consider a COR such as the collection of human beings composing our housing community and how it differs from a collection of physical human artifacts like desks. In both cases, the properties that define the class trace back to human beings; in the case of our housing community members, however, the class we're interested in is also a primary COR.

Likewise with other collective agents: the class of all mole rats has all mole rats as a primary COR; the class of all mammals ("mammals" as the literal taxonomic "class") has mammals as a primary COR. Likewise with individual people considered as a class of awareness occasions over time. Thus "Socrates" identifies the collection of all moments in his life, and it is to this same collection of moments that we turn (as far as possible) for understanding Socrates as an enduring, aware being. For all collective agents, we trace at least certain of the properties that define the class to the class itself.

Since we're interested here in the sources of properties that make us human and give us our way of being aware, what about matter and the cells of our bodies as primary CORs in this case?

To whatever extent the properties of our human way of being aware directly trace to the behavior of cells and matter, then cells and matter make up primary CORs. But as I already pointed out, our human way of being aware has its source not only in brain cells acting collectively but also in their collective behavior being sufficiently alike from person to person for humans to share cultural artifacts and, more broadly, all the realities that humans themselves create. If this is so, then those properties that make us human as distinct from other creatures cannot be traced to just the biology of individual cells. At least one of the primary CORs for these distinctly human properties must be composed of cells considered collectively as an embodied system that is a member of another collection; the COR must be composed of human beings as whole living creatures.

Lifelike Robots. Regardless of how alive or aware a physical object may seem, if it is not a member of a class that has itself as a primary COR, then the object is neither alive nor aware.

Imagine a robot so lifelike as to be indistinguishable in its behavior from a conscious human being (e.g., HAL in the movie *2001: A Space Odyssey* or Data in the TV series *Star Trek: The Next Generation*). It is tempting to ask what properties of consciousness would allow us to distinguish a being with consciousness (a human or hypothetically even a robot) from a lifelike robot without consciousness. In the context of this vocabulary, however, the important question is not *what* properties distinguish them but rather *where* the properties that distinguish them come from in the first place—which things of our world are the source of their own properties and which are not.[6]

Many complex artifacts—cars, computers, televisions, automated machines in factories—resemble living organisms in that the properties that distinguish them depend on many integrated components working together properly. Such a device's distinctive properties are only expressed when its parts are active and functioning as they should. When critical components fail, the device stops having these properties: the car no longer moves; the television no longer shows a picture. However, unlike living organisms, these distinctive properties originate not with the artifacts themselves but rather with human beings. It's humans that know a car as something that should move or a television as something that should show a picture; cars and televisions themselves don't know this. Recall in the chapter on Buddhism the example of an alarm clock going off when no one is around. Remove the human context and artifacts like these behave only as matter behaves; no other properties are necessary to describe their interactions, regardless of how complex or like a conscious human being the devices might be. None of their distinctive properties trace to a COR of which they are participating members.

Again, to distinguish things that are aware from things that are not, the important question is not *what* properties distinguish them but *where* their distinctive properties come from in the first place—what the primary COR(s) is/are for the properties governing their distinctive ways of interacting with the world.

Biological Cells. Robots may not form their own CORs, but biological cells do.

Consider the difference between biological cells and tiny robots. Like robots, the class properties of cells are expressions of cells as integrated, functioning wholes with parts that must work together properly in order for their distinctive properties to be expressed. Unlike robots, however, cells are what we think of as alive; they express properties that we trace to the cells themselves and that differ in certain respects from physical properties. As the basic structural and functional units of life, cells transform the rules of chemistry into the rules of biology.

In other words, we are back to where properties come from—to CORs. Cells are *themselves* collective agents for certain nonphysical publicities in our universe, whereas robots, insofar as they are strictly human artifacts constructed from matter, are not.[7]

As agents for the presencing of their own publicity, cells are aware per my working definition. But awareness as I've discussed it—and as we typically use the word—implies a first-person view, a subjective dimension. Why should we think that such terms as "first-person view" and "subjective" apply to cells? To use Thomas Nagel's touchstone for subjectivity, why should we believe that there is "something that it is like to be" a cell?[8]

Consider how the properties of a cell's environment—heat, pressure, and light, as well as the molecular properties of carbohydrates, proteins, and water—are all physical (as are the properties of what touches a person's skin, retina, ear drums, tongue, and nostrils). If cells are in fact a source of properties that are *not* physical, then cells *themselves* have the capacity to put the physical properties of their environment into *nonphysical classes.*

Consider what's involved in this accomplishment. The single-cell algae *Chlamydomonas reinhardtii*, for example, has the trait of moving toward light. In order to accomplish this behavior whenever it encounters light, the algae must be able to apprehend the light of one moment as an instance of light in general and the darkness of another moment as an instance of darkness in general. It must be able to process light as a *category* of experience tied to a certain here-and-now behavior of itself. No doubt we could program a robot to do this, but unless the robot was aware in the sense defined here, what it could *not* do is apprehend light in a way whereby the act of processing light as a category was *itself* an act of participation in a primary COR *responsible* for the category. At least part of the persisting meaning of the light to the algae comes from the algae, not from people, matter, or any other source.[9]

This persisting meaning over time, this value of light to an individual cell, is not only something that originates with the cells as living beings; it's also something we can't directly observe. When we look at a cell under a microscope, we don't see the value the cell itself finds in a light source; we see light as stimulus and the cell's movement toward light as response. Looking physically inside the cell also doesn't show us this value of light to the cell. Instead, we see DNA, RNA, proteins, amino acids, carbo-hydrates, and other molecules engaged in chemical reactions, all of which seem trace-able (theoretically at least) to the properties of matter alone. Just as we don't observe human first-person experience in the neurological behavior of a brain, nowhere in the cell's chemical reactions do we observe what light is like from the standpoint of the cell.

When we judge what in our universe has a first-person view and a subjective dimension, it seems to me that what we look for, at least in part, is this ability to put stimuli into categories having distinctive meanings for *specifically those exercising this ability*, distinctive meanings that cannot be directly observed from an outside vantage point.[10]

This example is especially important for illustrating my definition of awareness. At least part of what awareness entails here is the ability of living creatures to identify properties of their here and now that have meaning only in a context that's bigger than their here and now, *a context they also participate in maintaining*.[11]

The Term "*Chlamydomonas Reinhardtii*." Our words and concepts are human arti-facts and have groups of human beings as their primary CORs. Artifacts like these, however, differ from other artifacts by implicating a second sort of COR.

For example, the term *Chlamydomonas reinhardtii* refers to a certain kind of algae. The term is not the algae itself; it doesn't have the physical or biological properties of algae. However, it does refer to what has these properties. Our words and terms are human-specific realities whose COR is ourselves but that we produce to refer to other realities whose CORs need not be human at all. These elements of our speech and thought originate in groups of people joined by culture and language into what are sometimes called *communities of discourse* (CODs). CODs are those CORs responsi-ble for the words and concepts that serve as signifiers.

Thus words typically involve two varieties of CORs. On the one hand is the COD, which is the COR that produces the *signifier* (for us, a group of human beings). On the other hand are the CORs that produce what is *signified*, the collective agents for what our words refer to (e.g., actual algae cells).[12]

Because words don't need to have the same COR as what they signify, we're able to use words to refer to what is outside our direct experience (e.g., what it is like to be a bat, or a cell, or simply a person with different experiences) or what we are not expe-riencing at the moment (e.g., a toothache). This is also what allows us to use different words or glyphs in different languages to signify the same thing (e.g., "pen," "stylo," "कलम," "笔"), or to use different words or phrases in the same language to signify the same thing in different contexts (e.g., "morning star," "evening star," "Venus"), or to signify the same thing in terms of different properties (e.g., "animals with kidneys," "animals with hearts").

In our efforts to understand how our human awareness differs from that of other creatures, our extensive use of language is one place we might look first. With language would seem to come a new realm of reality, a realm inhabited by abstract artifacts of our own creation that provide extraordinary insights into realities other than those of our own creation.

Matter. As an agent of the universe, matter is its own COR (or CORs). But consider also how not just biological cells but also matter fits my definition for awareness. I've already described matter in general as a COR, and in fact, when we examine matter, we find it to be made of particles like electrons, quarks, and photons that respond to stimuli of the moment according to collective rules. It's to the collective behavior of these particles of matter that we trace publicities that we speak of as mass, gravity, momentum, energy, and so forth.[13]

Of course, we typically point to awareness not as a trait we have in common with matter but rather as one that distinguishes us from matter. I want to examine matter as a COR (or CORs) in some detail, and the next chapter is devoted to this and related subjects.

In the meantime, you might consider how a concept of matter that fits my definition of awareness might help with certain challenges in Western philosophy and science. If my definition of awareness in fact fits matter (i.e., if it offers a credible version of the philosophical view called "panpsychism"[14]), then wouldn't it be much easier to explain, for example, how life and human consciousness evolved from matter? Would it not also help with questions about how our first-person views and subjective experiences might be constructed out of cellular interactions that are, in turn, constructed out of atoms and chemical interactions? Furthermore, might it not also help with certain other problems like those of agency and causality (that I described in chapter 7 in the section "Individual and Collective Agency"), since they apply to matter as well as living beings?

Descartes Reconsidered

While we're on the subject of matter as a COR, let's briefly revisit Descartes.

Recall how I've reframed the Cartesian distinction between mind and matter so that it's not coupled with first-person versus third-person views. The mind-matter distinction is not one between how you or I see a house, person, or sun, for example, and the house, person, or sun as it "really is," independent of our minds. Instead, it's a distinction between two origins of the realities of our universe, both of which enter into making our universe what it is. Matter contributes its realities to the universe; our human mind contributes other realities.

In this context, the historical significance of Descartes's mind-matter distinction is the bright line he drew between two CORs—two CORs that proved to be of critical importance to our understanding of ourselves and the world we live in. More effectively than those before him, he disentangled the reality generated by our human minds from the reality generated by matter.

This is not to say that Descartes understood his distinction in this way; for him, reality continued to refer to what was independent of the human mind. And for those following him, this difference between mind and matter has been primarily a problem to be solved, as has been the case with our contemporary problem of consciousness.

This persistence of the mind-matter problem is highly significant in this context. I would argue that it resists resolution because it's of the nature of our universe for reality to originate within collections, within CORs, and that CORs arising from our human ways of being aware and CORs arising from matter are two kinds of CORs that have proved exceptionally important for us to distinguish.

Again, I want to make clear that reframing the Cartesian gap like this doesn't eliminate what I consider to be the more fundamental problem: the puzzle of how awareness can simultaneously play individual and collective roles. Instead, this reframing intends to disentangle the mind versus matter distinction (explained here in terms of different CORs) from what we experience as a difference between first-person and third-person views (explained here in terms of two roles of awareness).

Our Concept of Self in Space and Time

I want to close by returning to the subject of the self and personal identity—what it means to be an aware individual. I've discussed how publicities and their CORs occur in time as well as space. Notice, however, how differently we think of ourselves in the dimensions of time versus space.

In space, although my awareness of shared everyday realities relies on similarities in the way human beings are aware, I still think of myself as different from others. I am "here," and other people are "there." I have my habits, memories, and opinions; they have their own. In space, COR members are presented to each other as different individuals.

In time, however, when I think about all my past here-and-now moments of experience, I don't identify these experiences as ones that someone else had. I, *myself*, had these experiences; they are part of *my* life. In the dimension of time, my awareness of the world comes with a sense of *enduring* identity, with a sense of my persisting from one moment to the next. In the dimension of time, the COR members are presented as one self at different times.

In other words, my personal identity seems to align itself in space with my awareness's disjunctive role and in time with my awareness's conjunctive role. In space, awareness tends to identify its view of the world with a here-and-now occasion of conscious awareness that is individual and distinct from other occasions despite their similarities. In time, awareness tends to identify its view of the world with a here-and-now occasion of awareness that is collective and inclusive of other occasions despite their differences. Even though I am conscious only in my here and now, I nevertheless think of myself as being all my past experiences as well as my mostly unconscious shaping of these experiences into who or what I am now.[15]

This is an account of our everyday experience of ourselves. As I've discussed, who or what we might ultimately be is another question, one that the religions I've looked at answer in different ways according to their different—and, it seems to me, complementary—strategies for understanding reality. In the last chapter, I'll again take up this question of who or what we might ultimately be.

Summary

To facilitate my discussion of awareness, I've provided a working definition based on the two roles of awareness discussed in the previous chapter and the concept of everyday reality developed earlier (i.e., everyday reality as what combines publicity with the capability for presence). Awareness for my purposes here is the *presencing of publicity*—that is, it (1) establishes presence (its disjunctive role), (2) cooriginates/comaintains publicities (its conjunctive role), and (3) joins publicity with presence to produce everyday realities (both roles together). All three elements are interrelated; anything that acts in one of these ways is aware and acts in all of these ways.

When I say that awareness produces reality, I mean that it literally brings reality into existence. Awareness generates *both* what exists for us in our personal first-person views and what exists for us as the public world of our collective experience. The distinction that awareness itself makes between the standpoint of the individual and the standpoint of the collective underlies our conceptions of self, world, and deity, as well as our life goals and spiritual practices.

I've proposed that reality is anchored in many overlapping and nested collections of aware beings that include not only humans but also cellular life and matter, collections that I've called *collectivities of reference* (CORs). CORs are distinguished from *classes*: a class is the collection of a publicity's instances (all anthills, the desks in my house), while a COR gives an origin of that publicity (ants, human beings). One especially important subset of CORs, which I've called *communities of discourse* (CODs), serves us by generating abstract artifacts like words and concepts that can point beyond the CORs that originate them and provide extraordinary insights into realities other than those of our own creation.

The terminology used here reframes the Cartesian mind-matter distinction so that it's not one between an interior mind and an exterior reality but rather one between two different origins or CORs for reality—for what exists in our universe. Mind and matter may be very different, but they both generate realities; they both contribute to making our universe what it is. In this context, the historical significance of Descartes's mind-matter distinction is the bright line he drew between two CORs that proved to be of critical importance to our understanding of ourselves and the world we live in.

One of the most fundamental expressions of these collective activities of awareness is our concept of self, which derives, at least in part, from occasions of awareness that group differently in space and time. In space, we tend to identify ourselves with our local, here-and-now occasion of awareness (and awareness's disjunctive role); in time, we tend to identify with what endures from one occasion to the next (and awareness's conjunctive role).

CHAPTER ELEVEN

Physical Reality

Reality as Emergent

The terminology developed here addresses the Cartesian interior mind versus exterior matter problem by decoupling the mind-matter distinction from the "interior-exterior" (i.e., first-person versus third-person) distinction and treating both mind and matter as sources of third-person properties.

But this still leaves us with the question of how mind and matter relate to each other. Recall the problem of consciousness discussed in chapter 9. Even if we view mind and matter as two different sources of third-person properties, how do we reconcile our having minds and subjective experiences with our contemporary picture of the physical world?

Writers on this subject have offered two broad strategies: panpsychism and emergentism.[1]

Panpsychism, which I mentioned in the prior chapter, is the view that matter is ultimately of the same nature as our minds. According to this view, we have first-person experience because first-person experience is, in some sense, a feature of matter itself. This does not mean that particles of matter have minds like we have minds; rather, it means there is something about the nature of the mind or subjective experience or the first-person view that is basic to the stuff out of which our universe is made, including what we think of as "physical."

"Emergentism," on the other hand, is the view that properties like mind and consciousness arise out of and depend on matter but do not themselves characterize matter. By this account, a collection of entities (e.g., cells, subatomic particles) can together generate properties that cannot be found in the properties of the individual members of these collections. This idea that collective behavior can exhibit distinctive properties not evident in the behavior of smaller-scale individuals is not at all new. Aristotle, for example, spoke of the whole being greater than the sum of its parts, and John Stuart Mill (1806–1873) wrote on the subject. Historically, emergent theorists have supported their thesis by pointing to biological phenomena like ant colonies, which are capable of behavior that is much more adaptive and intelligent than individual ants are by themselves. They also point to evolution, whereby individual one-celled organisms have evolved over time into colonies that eventually became multicellular organisms such as ourselves with properties not evident in the simpler life-forms.[2]

Emergence as it is discussed today is usually divided into two kinds, both of which are important in my context here. Recall the part-whole construction of particulars described in chapter 3.[3] Both kinds of emergence seek to explain how wholes arise and exhibit properties not found in the properties of the parts. For example, the physical property of "heat" doesn't exist for individual particles of matter like atoms. It describes the collective effect of many interacting particles, a property that emerges when small-scale particles of matter interact with much larger-scale beings like ourselves who experience these particles in aggregate rather than as individuals.

This example of heat describes emergence in its "weak" sense (sometimes called "epistemological" emergence). Here, while the collective effect of individuals (e.g., atoms in aggregate) creates novel properties not found in individuals themselves, these properties still can be traced (e.g., via a computer simulation) to the properties of interacting individuals.

Emergence in its more controversial "strong" sense (sometimes called "ontological" emergence) refers to properties of wholes that cannot be traced, even in principle, to the parts making up the wholes. These are properties that some believe can influence the properties of the parts from which they've emerged (that have "downward" causality). "Life" and "consciousness" are often thought to be emergent properties in this stronger sense.[4]

The distinction between weak and strong emergence offers another way to understand the terminology proposed in this book. Consider how realities are generated and maintained by collectivities of reference, or CORs. Weak emergence pertains to realities (e.g., heat) that arise from the interactions of members of *existing CORs at different scales*. Human beings relate to matter not at the scale of atoms and photons but rather at the scale of matter's collective behavior. We don't perceive atoms themselves; we perceive heat, mass, pressure, and the like—what atoms produce in aggregate. Strong emergence, on the other hand, pertains to the generation of *new CORs*.

In this case, the collective behavior of a COR at one scale (e.g., composed of matter) gives rise to another COR at a larger scale that is different from itself (e.g., composed of living cells), a COR that produces its own realities.[5]

Note here how even strong emergence is the origin not of awareness as defined here but rather of different *kinds* of awareness. Both panpsychic and emergent theories are combined within my working definition of awareness. To define awareness in a way that applies to matter as well as ourselves is to propose a panpsychic answer to the mind-matter question. But the definition also proposes that awareness comes in kinds that vary greatly according to the publicities presenced, especially publicities presenced at different scales. Awareness as I've defined it intends to name that feature of the universe whereby emergence occurs. Awareness, to repeat, is not so much a property *of* our universe as that whereby properties arise in the first place.[6]

Let me review how the vocabulary I've proposed melds both panpsychic and emergence perspectives.

Beginning with matter, its quantum mechanical behavior originates publicities at one (or, more likely, several) extremely small scales. This variety of awareness (as I've defined it here) generates properties such as mass, gravity, and electromagnetism that are so unlike the publicities we ourselves generate as to raise the question of whether the ability to generate publicities at these scales should be called "awareness," despite it fitting my definition for the term.

Then, at a much larger scale, biological cells use these publicities of matter, these physical regularities, to create an entirely different set of publicities. This is the level of cellular life and DNA and a very different kind of awareness. At this level, awareness produces all the regularities of nature we associate with living beings. They reproduce, communicate with each other, defend themselves, nourish themselves, and excrete waste. These additional properties originate not with the behavior of individual subatomic particles or their groupings in atoms or molecules but rather with the behavior of many trillions of atoms networked together and acting as one being—one cohesive group or society.[7] When living cells reproduce by splitting in two, it's not quarks or atoms that reproduce, that make copies of themselves; it's *systems* of atoms that duplicate themselves. Then, too, when single cells talk to each other (which they apparently do almost all the time when not removed from their natural circumstances[8]), this chemical communication is not between subatomic particles or atoms. It is, again, between *groups* or *systems* of atoms integrated together as whole biological entities.

At the still larger scale where we ourselves live and breathe, animals use the publicities of cell behavior to generate still another, different category of publicities—those of multicellular organisms. Again, we have a different kind of awareness, which now in many cases seems somewhat like our own. This time, reproduction does not mean individual cells making other individual cells; it means organizations of trillions of cells making other similar organizations of trillions of cells. This time, the "talk" between beings takes place not quark to quark or cell to cell but rather multicellular organism to multicellular organism. The individual neurons in my brain don't talk to the individual neurons in another person's brain; a great many neurons acting together in my brain talk with a great many neurons acting together in the other person's brain.[9]

Mathematics and the Behavior of the Universe

How can it be that mathematics, being after all a product of human thought which is independent of experience, is so admirably appropriate to the objects of reality?

Albert Einstein[10]

On July 4, 2012, scientists at CERN in Geneva, Switzerland, announced that their Large Hadron Collider had found traces of a long-sought subatomic particle, the Higgs boson. Prior to this discovery, the Higgs boson was just a mathematical forecast, an idea invented to fill out our favored model of physical reality. Finding evidence of this particle's existence was yet one more example of the peculiar power of mathematics not only to describe reality but to make predictions about its composition and behavior.

There are many such examples. In 1959, researchers found empirical evidence of the ethereal subatomic particle, the neutrino. That particle, too, had been merely a concept invented by physicists—"an accounting device, a mathematical fiction to make a set of equations fly," as George Johnson put it in a 1995 *New York Times* article.[11]

In that same article, Johnson went on to tell yet another story in which "theorists needed a new particle to satisfy their sense of aesthetics, and nature went on to oblige":

> In the early 1930's Paul Dirac, the English physicist, was staring at the equation he had divined for explaining the electron when he noticed that it could be solved two ways. The square root of 4 is either 2 or −2. Similarly, one solution of Dirac's equation yielded the negatively charged electron, just as he had intended, but the other yielded a particle no one had ever seen—one that was exactly like the electron except with a positive charge.

Even Dirac seemed reluctant to believe that Nature took his equations quite so seriously as he did. But soon after, the track of one of these antimatter positrons (the mirror image of the signature left by an electron) was discovered in a detector called a cloud chamber. "The equation," Dirac exclaimed, "was smarter than I was."[12]

This "unreasonable effectiveness of mathematics in the natural sciences" (as physicist Eugene Wigner once described it[13]) seems to make sense only if our universe behaves mathematically, at least in our awareness of it. But why should this be so? Why should something so abstract and seemingly conceptual as mathematics so effectively predict the behavior of nature?

Recall how this and prior chapters have described the relationship between individuals and their collections as fundamental to the nature of reality. Next, consider one of the major historical milestones in mathematics, the development in the nineteenth century of *set theory*.

Set theory is founded on a single basic relationship, that between a collection and its members—in other words, the same individual-collective relationship that I've described as underlying reality. In its most primitive form, a *set* is simply a collection of things: a species of tree, for example, or a kind of subatomic particle or a group of numbers or even a collection of sets. The collection can be determined by a list—for example, the set of the letters Z, B, and X. Or it can be determined by properties—for example, the set of all atoms with one proton in its nucleus. The general idea of sets has been around since humans have grouped phenomena in categories according to various properties. Any universal describes a class of instances, which is close to the modern meaning of *set*.[14]

Set theory's relevance to my discussion here should already be somewhat clear. It traces out in the language of formal logic the implications of the individual-collective relationship that I've associated with the nature of reality. But it's a subsequent mathematical discovery that makes set theory especially relevant.

Following set theory's initial development, mathematicians found that *literally all of mathematics*, even numbers themselves, could be derived from set theory, as could much (though, importantly, not all[15]) of logic. Saying that something behaves mathematically is the same as saying that its behavior is characterized by the individual-collective relationship and its implications.[16]

The concept of reality I've been developing here also involves collections of individuals; it traces what we call "real" to collectivities of reference. If CORs are in fact a defining aspect of reality, then there is no mystery why our universe seems to behave mathematically. Mathematical (i.e., set-theoretical) behavior is entailed simply by something being "real"—by it being part of a third-person world distinguished by its public (i.e., collective) nature.

Moreover, the definition proposed here for "awareness" suggests a place to look for the source of set-theoretical behavior in our universe, a place we might find the agency of our universe's mathematically describable orderliness and logic. We as aware beings are agents of logic for those realities of the universe that originate with us; other kinds of awareness—other ways for publicity to be presenced—are agents of logic for the realities that originate with them.

Also important in this context, set theory in its most primitive form (or "naive" form, as it's usually put[17]) *generates paradoxes.* One of the basic characteristics of sets is that the members of sets can be themselves sets. Thus the set of "all trees" has, for example, "chestnut trees," "maple trees," and "oak trees" as members, which are each themselves a set containing different species of the tree. If, however, we permit sets to contain *themselves* as members (e.g., the set of all abstract concepts is itself an abstract concept and thus should be a member of itself), it can be shown that certain sets both do and do not contain themselves.

Perhaps the best-known example of such a paradox comes from Bertrand Russell. Suppose we restrict sets to just those that do not contain themselves. In other words, we define a set composed of all the sets that do not have themselves as members (perhaps in an attempt to eliminate paradoxes). Russell noticed that if we now ask whether this parent set is a member of itself, we find that it is paradoxically both a member and not a member of itself. If it is a member of itself, then it's not a member of itself; if it isn't a member of itself, then it is.

Russell's paradox is often illustrated by the example of a barber who cuts the hair of everyone in his town who doesn't cut his or her own hair. But what, then, of the barber? Does the barber cut his own hair? If he does, then he doesn't; if he doesn't, then he does.[18] (Notice that the paradoxes of set theory arise because, as I've been discussing, the individual-collective relationship is *inherently* enigmatic in this way.)

I have one final comment on the paradoxes at the core of set theory and thus of mathematics. In another of mathematics' great milestones, Kurt Gödel demonstrated in 1931 that it was impossible to create a system of arithmetic (apart from trivial, meaningless ones) that could prove all the statements that were true within that system. One could always derive statements that the system could neither prove nor disprove.[19] While Gödel's proof applies only to formal arithmetic systems, it strongly suggests that what we generally mean by "true" is something more than or other than "logically provable." We are capable of recognizing a statement as true while at the same time recognizing it as unprovable (or meaningless or paradoxical).

Two Pictures of Physical Reality

From mathematics, let's move now to the physical aspects of our universe that mathematics seems to describe so well. I have claimed that matter fits my working definition for awareness, but how well does it fit? In what specific ways might physical behavior be consistent with awareness's curious combination of conjunctive and disjunctive behaviors?

I'll start with a story from science about the early debate over whether light was wavelike or particle in nature. It was a very important debate in that its resolution in the early 1900s by Albert Einstein set the stage for quantum mechanics and our modern understanding of physical reality.

Toward the end of the seventeenth century, the prevailing theory was that light was made of small particles, or "corpuscles," traveling through space like minuscule billiard balls. Then, in 1690, Christian Huygens came up with another theory. He proposed that light could be explained as a wave oscillating in an invisible medium called "ether." Unfortunately for Huygens, his contemporary, Sir Isaac Newton, sided with the corpuscles version. Without adequate evidence to settle the dispute, Newton's prestige won his version greater favor for a century.[20]

Then, in the 1800s, experimental evidence started coming in against Newton's version. If, for example, light were made of particles as Newton said, then the image light made on a white surface after passing through two side-by-side narrow slits should show two side-by-side bands of light (something like the pattern a lot of bullets would make if they were shot from the same place through two side-by-side narrow openings). But when Thomas Young performed this experiment in the first decade of that century, what he saw instead was a wavelike interference pattern much like the one water waves made in analogous conditions (see figure 7). Later, in 1864, James Clerk Maxwell's electromagnetic account of light and the experimental confirmation of that theory by Heinrich Hertz in 1886 gave the wave theory even more support.

But the wave theory failed to explain other phenomena (e.g., the "photoelectric effect").[21] It was Einstein in 1905 who solved the problem by proposing that light and other forms of electromagnetic radiation were waves made up of tiny bundles, or "quanta," of energy that we now call "photons."

Einstein's solution revolutionized the way we thought of light and subsequently all material reality. Light, it turned out, could not be characterized as particle alone or wave alone; it was both. Under some circumstances, it presented itself as a particle; under other circumstances, what we saw were waves.

In fact, two very different pictures of matter emerged in the previous century, one based on particles (the matrix mechanics of Werner Heisenberg) and one based on

FIGURE 7. Wave pattern made by light projected through two side-by-side narrow slits (illustrated). Drawing by CranCentral Graphics.

waves (the wave mechanics of Erwin Schrödinger). While these two pictures were quickly shown to be equivalent, the way matter appears to us in each picture is very different and, taken together, would seem quite at odds with the way we usually think of reality. After all, how can something be both particles *and* waves?

Consider, however, this dual behavior of matter in my context here. Light as particles is light wherein its collective nature recedes such that what is presented is the particle behavior of individual, interacting quanta of energy. Light as a wave is light wherein its particle nature recedes such that what is presented is collective behavior as a continuous thing in itself. In other words, twentieth-century physics has given us two pictures of physical reality that can be traced to a distinction between individuals and their collections—the same distinction that I've proposed underlies many of our philosophical problems and that I've incorporated into my definition of awareness. In the case of matter, the individual appears as point particles (or perhaps "strings"), while collections appear as a seamless continuity.

I want to close this section with a fairly long quote from Hans Christian van Baeyer, who used to write a column on physics for a journal called *The Sciences*. The interplay between individuals as point particles and the collective as a seamless continuity has been a long-standing central theme in our changing characterizations of matter. This passage describes how we've historically cycled back and forth between understanding matter as "points" and understanding it as a "continuum":

Points or continuum? Integers or real numbers? Discreteness or connectedness? How is the world constructed? How should it be described? The struggle between those opposing views is almost as old as physics, and it shows no sign of abating. In fact, many of the major milestones in the history of physics—and particularly in the history of the understanding of matter—were accompanied by a switch from one side to the other in the debate.

Early Greek philosophers taught that all matter is made up of one or more continuous substances: earth, water, air and fire. The atomic hypothesis, which substituted discrete point particles for continuous substances, was motivated in part by dissatisfaction with continua.

Fast-forward to 1926. Electrons are regarded as point particles orbiting a nucleus. From time to time they suddenly change orbit and emit or absorb a photon with a discrete frequency. The Austrian physicist Erwin Schrödinger decides to describe the process in the manner to which he is accustomed from classical physics, which deals with continuous motions through the space-time continuum. But the electrons apparently move discontinuously. As a way out of his dilemma he invents what he calls an "emergency exit": wave mechanics. The makeshift element of the theory is a continuous wave function, which is not a real object at all, but merely encodes information about probable outcomes of experiments. Schrödinger hinted that what he felt to be unsatisfactory about his own theory might be related to the mismatch between reality and the artificial continuum of numbers.

In time, wave mechanics yields to quantum field theory, which fills the world with swirling, overlapping continua—one continuum for every elementary particle. . . . Then quarks were discovered, and the image of the world was once more repainted in the pointillist manner. Now points are being replaced by continuous strings.

But the last word has not been spoken.[22]

These days, for example, there are efforts to explain the arising of space-time in terms of the entangled interactions of tiny bits of information—the "it from qubit" theory.[23]

Quantum Strangeness

There is more to the story of these two pictures of matter. I've pointed out that the relationship between individuals and collections is in many ways mysterious and, when treated as a fundamental principle, causes paradoxes (e.g., those of set theory). In physics, we observe particles of matter behaving in these contrasting individual and collective ways, *even when* such behavior seems to defy conventional logic.

Let's return to the double-slit experiment that gave support to the wave theory of light in the early 1800s. In 1909, four years after Einstein proposed that light was both wave and particle in nature, G. I. Taylor repeated Young's double-slit experiment, but this time with light so faint that only one particle of light could logically be going through the slits at any one time. Since there was nothing left to interact with, one might expect the wave pattern on the shadow side to revert to two white bands with no interference pattern. But this is not what happened. In fact, Taylor's double-slit experiment was the first of a number of experiments in the previous century that demonstrated that material reality was much stranger than anyone had imagined not so many years before.

First, Taylor discovered that although only one photon could be going through the slits at a time, the pattern on the photographic plate showed the *same* interference pattern as before. Even though there were no other photons to interact with, the photon still behaved as a wave. It continued to exhibit the behavior of a seamless continuity even when all that remained of the continuity as it passed through the slits was a single photon.

Second, the same results showed up for other particles of matter. Beams of electrons (and later beams of atoms and even molecules) produced interference patterns, and they did so even when passing through the apparatus one at a time. All matter seemed to exhibit this dual individual-collective behavior.

Third and most bizarre, it was shown mathematically and then experimentally that if experimenters set up their equipment to record the hole through which individual particles actually passed, the interference pattern disappeared and became just *two bands of light*. Any interaction with the particles that revealed their individual behavior would remove (or "collapse," as it's usually put) the wavelike behavior they normally exhibited.

This double-slit behavior is remarkable not only in itself but also because of what it tells us about matter as we observe it at these very small scales. Since it is impossible to discover which hole particles pass through without collapsing the wave, the *reality* of such a wave insofar as any observer is concerned is inherently an expression of the particles' indefinite positions. As long as observers don't interact with the particles themselves, the wave presents itself not as a collection of discrete particles of matter with well-defined locations but rather as a pattern of *probabilities* for particle locations. A beam of light (or of electrons or atoms or molecules) travels as a "probability wave," as it's sometimes called.[24]

This experiment has been repeated many times with the same results. Interact with beams of photons so as to observe which hole they pass through, and they behave as individual particles. Stop interacting in this way, and they behave as wavelike possibilities for interaction.

Richard Feynman, one of the great teachers of physics and a Nobel Prize winner in his own right, regarded the double-slit experiment as central for understanding quantum mechanics because it revealed, in his words, "a phenomenon which is impossible, *absolutely* impossible, to explain in any classical way, and which has in it the heart of quantum mechanics. In reality, it contains the *only* mystery . . . the basic peculiarities of all quantum mechanics."[25]

Yet as you consider the results of these experiments, consider also my characterization of our own everyday reality as a composite of many, often competing collective *capabilities* for presence and how here-and-now interactions *select* among these options. Thus the people in the houses around me can be doing any of many things right now, but because I'm not interacting with them now and don't know what they are doing, it's only the probabilities that exist for me, not their hidden-from-me activities. Any choices I make relating to them will be based on these probabilities—what I know about them right now. If, however, I talk to or otherwise interact with them, then that activity selects among these collective capabilities for presence. The probabilities give way—"collapse"—into the currently expressed properties.[26]

It's important to understand what I'm *not* saying here. I'm not saying that what really exists are these hidden-from-me activities and that probabilities come into play merely because I don't know what is actually going on. I'm saying the opposite—the *probabilities* are the reality. What is real is quite literally composed of the possibilities for awareness. The reality of what is hidden is not something apart from its hiddenness.

I'm also not saying that what really exists is just what the universe looks like from the perspective of individuals like you or me or particles of matter. The reality of interest in this book is the third-person view, what is public, how things are for collections of individuals. Regardless of whether we're examining the behavior of human beings or matter or whether probabilities collapse or not, reality in the sense I mean it here refers to the behavior of that which presents itself to us as a shared universe.

FIGURE 8. Marcel Duchamp's 1916 *With Hidden Noise* offers an artwork treatise on the reality of mystery. Concealed within a ball of twine sandwiched between two screwed-together brass plates is an unknown object added by Walter Arensberg that makes a rattling sound when the sculpture is shaken. What is real for us is not whatever the concealed object is were it known to us but rather the object as something unknown—a "hidden noise." © Succession Marcel Duchamp / ADAGP, Paris / Artists Rights Society (ARS), New York, 2017. Photo courtesy of Philadelphia Museum of Art (#1950-134-71).

Apart from these parallels between human and material behavior, it's not clear how far these similarities extend in this individual-collective regard. There are obvious vast differences between human awareness and whatever might constitute "material awareness," but exactly where the parallels end and the differences begin seems to me more a question for physicists and cognitive scientists than for philosophers.

It's also not clear where material equivalences might lie for what we at our scale of existence think of as aware beings. Are atoms aware individuals per my definition of awareness? Do subatomic particles—bosons and fermions—in fact fulfill my definition for awareness? Might the loci of material awareness lie elsewhere (as well or instead), especially considering some of the ways material particles coordinate their behavior over space and time (as when particles are "entangled")? These also seem to be questions better answered by physicists than by philosophers.

What I do say is that matter at its most fundamental level seems to share basic traits with what we would usually think of as aware beings, traits I've attempted to capture in my definition of awareness. What I call "awareness" seems to be characteristic of not only living beings but also matter.

Schrödinger's Cat

These similarities between what we observe at our scale and what we observe at much smaller scales show up in other ways as well. Here is another of the great curiosities of quantum mechanics, the one called "Schrödinger's cat."

Recall that Erwin Schrödinger was the one who became associated with matter being pictured as a wave. He was also a colleague of Einstein, and they both were upset about the bizarre implications of quantum mechanics. To illustrate this strange behavior, Schrödinger imagined the following situation.

Picture a closed box with a radioactive atom that has a fifty-fifty chance of losing part of its nucleus—"decaying"—in the next hour. Also in the box is a device to sense if the atom decays, a vial of poison gas, a cat, and a mechanism to smash the vial if the atom decays. If the atom decays, the vial of poison breaks and the cat dies. If the atom doesn't decay, the vial of poison remains intact and the cat lives. All this happens hidden from view inside the box.

In our everyday practical sense of thinking about such things, we would say that at the end of the hour there will be either a live cat or a dead cat in the box, but certainly not both. What bothered Einstein and Schrödinger was that, in the quantum world, probabilities were the reality, and the cat should be neither alive nor dead until we looked at it. Until we opened the box, the atom would have neither decayed nor not decayed. Both states would be superposed; both would be reality until we opened the box.

We have since learned that macroscale objects such as a cat cannot be linked—"entangled"—for a significant length of time with microscale quantum events such as a decaying atom. At the end of the hour, the cat is either alive or dead but not both.[27]

Nevertheless, insofar as concerns just microscale phenomena, superposed states are normal. In experiment after experiment, we have found that a material state can both exist and not exist at the same time. It has become such a physical given that we've been experimenting with quantum computing, in which the storage unit has three states instead of two. Instead of just "on" or "off," it can be "on," "off," or various degrees of "maybe."[28]

Most people have found this superposition of states to be very strange. How can a thing both exist and not exist at the same time? Consider, however, my characterization of our world as a composite (or "superposition") of what is capable of public presence for different CORs. In this context, Schrödinger's cat describes a situation where matter provides one reality, the cat another reality, and humans yet a third reality. If the matter and the cat realities are hidden from humans such that all we know about the cat is that there's a fifty-fifty chance of it being alive, then *isn't that fifty-fifty chance in fact our reality?* And in regards to what we've learned about entanglement since Schrödinger first imagined his cat, doesn't the cat's more decisive situation describe the state of the cat *from the perspective of the cat* and its own interaction with the world?[29]

My point is that this composite reality is not just the reality of subatomic particles; it's our own everyday reality, as well as the everyday reality of any creature that is aware in the way I've described. If I go into an unfamiliar, dark room and grope for a light switch, part of the reality of the light switch is certainly its exact physical location—what is maintained by matter. But if I'm in the dark and don't know its exact location, then how can that exact location be the only factor governing how I and the switch interact? Reality surely has to do with these possibilities for interaction, which, in turn, depend on different collectivities of reference.

It seems to me that, specifically in regards to superposed states, the reality of subatomic particles is no more bizarre than our own reality *as long as* we think of these particles' interactions as awareness events and of reality as originating with CORs. Schrödinger's cat presents a paradox only if we consider the life-death status of the cat in the box as the same for all collectivities of reference.[30] (This does not mean that I think the terminology proposed here demystifies other aspects of quantum mechanics, such as entanglement. In fact, I think that such peculiarities of matter's conjunctive behavior indicate that awareness in general has depths that we have barely begun to understand.)

I should again stress that while I believe the terminology proposed here will help considerably with certain issues related to the nature of reality, it does so more by moving these issues to another venue than by resolving them in any final way. Schrödinger's cat may not seem especially paradoxical if reality is a composite of properties traced to different CORs. But it seems to me that reality thought of like this opens up a new universe, a metadomain of reality. What we once thought of as universally true for all now becomes what is true for different groups of individuals, and the "universally true" becomes a strange, often contradictory amalgam of how all these subtruths might fit together—the fugue of this book's title.[31]

REALITY'S FUGUE

Summary

Philosophers have offered two broad philosophical strategies to reconcile mind and matter. Panpsychism proposes that matter is ultimately of the same nature as our minds; we have first-person experience because first-person experience is, in some sense, an aspect of matter itself. "Emergence" proposes that properties such as mind and consciousness arise out of and depend on matter but do not themselves characterize matter. By this account, a collection of entities (e.g., cells, atoms) can together generate properties that cannot be found in or traced to the properties of the individual members of these collections.

The terminology proposed here combines both panpsychic and emergence approaches; awareness names that feature of the universe whereby emergence occurs. Awareness is not so much a property *of* our universe as that whereby properties arise in the first place.

As the agency of the individual-collective relationship in our universe, awareness as defined here also intends to explain why the universe behaves mathematically. All of math is based on the individual-collective relationship of set theory. Awareness points to the agency of set-theoretical behavior in our universe.

The distinction between individual and collective standpoints has been central to my analysis of awareness as well as to my analysis of conflicting strategies for understanding reality. I've also illustrated how humans relate to each other and to the world from these two standpoints. In physics, we find that material particles also relate to their circumstances in these two ways. One way is based on particles (matter from the standpoint of the individual); the other way is based on waves that are collectively coordinated possibilities for interaction (matter from the standpoint of the collective). Moreover, the paradox of quantum mechanics "Schrödinger's cat" illustrates how matter can exist in two opposing states at the same time. As strange as this may seem, as long as no other peculiarities of quantum behavior are involved, it is consistent with reality thought of as a composite of possible properties (i.e., publicities with various capabilities for presence) generated by different CORs.

CHAPTER TWELVE

Religions Revisited

Philosophy, Religion, and Mystery

I've proposed understanding conflicting accounts of reality as a fugue of contrasting themes. I've further proposed that what binds these themes together into a single composition is a philosophical puzzle about individuals and their collections that is impervious to contradiction-free resolution, a puzzle that each theme approaches differently. It's a puzzle that's basic to the nature of our universe as we discover it in our perceptual experience and our reasoning about that experience.

To clarify this predicament, I've contrasted the strengths and weaknesses in three accounts of reality found in world religions, which, however, may have left the impression that these accounts were philosophically incomplete in some way. Their metaphysical approaches neglected a fundamental piece of the individual-collective puzzle.

The puzzle itself seems to imply this conclusion. While an account of reality anchored in both first-person and third-person views might best match our everyday experience, constructing a comprehensive and coherent version of such an account seems to be like trying to build a model of reality on the edge of a knife—apply any logical strategy for bridging the gap between what is present in individual experience and what is public for collections of individuals and our efforts seem to tumble to one side or the other. Either we get a third-person account that attempts to explain individual experience in terms of publicities or we get a first-person account that attempts to explain publicities in terms of individual experience.

So how are we to overcome this philosophical hurdle? Is it even possible to construct a philosophy that binds together both pieces of the puzzle in a logically coherent way?

The religions discussed earlier are of considerable interest in this regard. Despite whatever philosophical difficulties they each might have, all of them have developed strategies for overcoming this individual-collective obstacle. These strategies involve at least two elements.

First, in marked contrast to much of Western philosophy, all these world religions make room for *mystery* at the heart of their metaphysics.[1] This does not mean that their philosophies abandon reason; rather, they presume that reality holds a puzzle that cannot be solved in an ordinary way.

Each of these religions provides a sophisticated account of reality consistent with its underlying assumptions about where such an account should begin. However, each religion also makes a sharp distinction between this conventional understanding of reality and another sort of understanding that transcends the conventional—one typically represented by enlightenment or by God. Simply being a devout Buddhist or Hindu doesn't mean that one is enlightened.[2] Simply being a devout Jew, Christian, or Muslim does not mean that one's perspective on reality and truth is similar to God's. In all these religions, one does not automatically understand the true nature of reality simply by being a participant in the religion and understanding its philosophical view. The ultimate truth of reality transcends our customary understanding. It's not accessible through conventional avenues of science or philosophy.

Second, while participation in one of these religions may not automatically grant understanding of ultimate reality, it does intend to aim you in that direction. Each religion lays out a path toward wholeness, a way of embracing that piece of the philosophical puzzle that is obscured by the religion's own philosophical beginning and that is responsible for reality appearing mysterious in some way.

I've made this point in prior chapters, but let me repeat it here. It has to do with who we are as human beings and what religions in general say our lives are ultimately about.

Let's begin with Nondual Hindus. Here Brahman is characterized as the universal aware self animating our cosmos and, as such, is beyond what can be named or described. But Brahman's mysterious nature does not mean that our identity with Brahman is beyond grasping altogether. Nondual Hinduism is fundamentally a path toward moksha, or liberation, and toward experiencing our underlying identity with Brahman.

Significantly, this understanding we seek comes *only in direct here-and-now awareness*. Even though one may begin by tracking down that which is most universal, one's goal in Nondual Hinduism is to find this universal source of all things within the *presence of one's own here and now*. Although one begins with a concept of ultimate reality shaped by the standpoint of the collective, one achieves enlightenment only when the standpoint of the collective merges with the standpoint of the individual—only when that which is most public becomes present.

Buddhism, on the other hand, typically deemphasizes universals in its accounts of ultimate reality in favor of what is present in direct awareness—what the world looks like from the standpoint of the individual. Such a beginning does not deny the existence of universals in our lives but sees them as originating in collections of interacting aware beings. In Buddhism, it is out of such interactions that the universe arises. Out of *my* presence's interdependent relationship with *your* presence, *our public* universe emerges. The goal for Buddhist enlightenment is thus the inverse of the Hindu one. Instead of beginning with what is most public and advancing toward its presence in here-and-now experience, an aspirant begins with what is most present and advances toward its expression as a genuinely public world. One begins with the interactive dynamics of one's here-and-now experience and advances toward discovering the public universe as arising from within that presence.

As in Nondual Hinduism, therefore, one achieves enlightenment only when one's own standpoint of the individual merges with the standpoint of the collective—when what is present in here-and-now experience becomes the arising of what is public for all of us. And this interdependent arising of a public world is, you might recall, one of the very few uncontested, genuine universals in all of Buddhism.[3]

In contrast to these Asian traditions, Abrahamic monotheism conceives of our world as having two sources. God is the collective agent behind the rules governing the way events unfold, while we—ensouled beings—are agents of individual cause and effect subject to these collective rules. The problem here is not that the first principles of these religions' metaphysics subordinate one piece of the puzzle but that they need to fit the pieces together in some logical way. And invariably, it seems, the two pieces simply don't fit. Try as one might, one still ends up with two pieces. As long as we think of ourselves as individual agents only, we are left with a philosophically troublesome gap between ourselves and the collective agency of God.

Yet although the relationship between these two realms of reality is mysterious, these monotheistic religions ask us to unite them as best we can. The goal in life is to overcome as far as possible our separation from God. We are to try to perceive the world as God might perceive it, love all creation as God might love it, and act in the world as God might have us act. And if we look beyond religious orthodoxies to these religions' mystical traditions, we find more esoteric teachings that point us toward a more complete union with God.

In sum, each of these religious philosophies is complete, a unity that deals with both public and present aspects of reality. The differences occur in how these aspects are integrated—what is to be highlighted in the conventional orthodoxy and what is to be a mystery.

If these accounts of reality are complete in this way, then their differing points of view are not necessarily flaws that can be overcome by more comprehensive philosophical accounts. Rather, they are more the faces that these religions present to the world, their personalities as distinct from the personalities of other religions and philosophies. This certainly doesn't mean that participants in these religions understand their metaphysics as one point of view out of many. Narrow-mindedness and intolerance of other views are legion in religions. It simply means that religions themselves differ one from the other much like people differ one from the other, each having their own distinct worldviews. The naivety of certain religious adherents should not be presumed of the cosmologies to which they adhere.

There is another, important way the philosophies of the religions I've discussed here differ from at least some contemporary philosophical views.

In our quest for universal truths, no doubt the distinction between first-person and third-person views has been and remains central. It contrasts what is true for individuals with what is true for groups (or all groups generally), irrespective of individual points of view. But the distinction between these two views doesn't entail these views having no relationship with each other whatsoever. It doesn't imply that awareness lies strictly on the first-person side and that the gap between first-person awareness and third-person reality is so absolute that we can conclude on this basis that awareness in our universe is somehow manufactured by a reality without awareness. None of the religions I've highlighted here supports such a position; in none of them does awareness arise from something that is not aware.[4]

It is rather the reverse of this. Nondual Hinduism begins with the first-person view and takes the aware self as its first principle; from this, all the universals and particulars of our universe arise. Buddhism also starts with the first-person view and awareness, but in a distributive rather than a collective sense. It emphasizes an individual's own here-and-now experience and conceives of reality as arising out of our everyday first-person interactions. As for various theisms, to trace the agency of collective phenomena to one or more deities is likewise to put the first-person view and awareness center stage, albeit as characterizing a deity. From a philosophical perspective, a deity may be conceived of as embodying a third-person, collective standpoint. To participants in a theist religion, this embodiment typically manifests in the form of an aware being of sorts, an entity to which one can pray and who is influenced by ritual (*despite* the difficulties in rationalizing this view).

I and these religions may, of course, be wrong about the centrality of awareness in the cosmos. Nevertheless, something at least similar to what the term awareness refers to seems to lie at the heart of many of our most difficult philosophical problems. Granted, we're left with awareness's perplexing nature, but awareness seems to pose problems regardless of philosophical account.

A Riddle Blocks Our Way

Before looking further at what religions say about moving past this individual-collective puzzle, I want to be clear about how thoroughly intractable this puzzle is in terms of conventional logic alone. We are like Oedipus trying to get to Thebes and forced by the Sphinx to answer a riddle[5] on which our lives depend, except that in our case, the riddle seems impossible to solve in any customary way. Its solution seems to have less to do with discovering the correct answer than with discovering the nature of the riddle itself.

The philosophical barrier raised by this riddle should already be evident from the many long-standing philosophical problems that I've traced to individuals and their collections. Of these problems, I've discussed contradicting accounts of reality the most; let me briefly review some of the other difficulties.

One of the oldest and most important problems discussed in this context is that of universals and their relationship to particulars. While our universe appears to be made out of particulars characterized by properties that remain the same over collections of particulars, it has been extremely difficult to understand how particulars relate to their collective characteristics. Do universals reside in particulars as Aristotle said, or in a realm of their own per Plato, or rather in our minds? How do universals come to characterize particulars? What would be left of a thing to make it a particular were we to strip away all the universals that characterized it?[6]

This book reframes these questions in terms of publicities and presence, but the questions don't go away; they just change. My terminology intends not to solve these problems but rather to consolidate them with other philosophical problems. On the one hand are publicities, which encompass both universals and particulars.[7] On the other hand is presence, which is that whereby publicities participate in here-and-now interactions—that whereby all the possibilities of this moment come to be for me this desk here or that bird outside my window there. The puzzle now becomes one of how individual here and nows might presence publicity, a process that I've associated with what we usually call "awareness" insofar as it relates to ourselves.

It's this publicity-presencing process ("awareness," or whatever name you prefer) that I find behind the individual-collective puzzle and at the epicenter of many philosophical problems.

Consider, for example, the perplexing way reality presents itself to us in time. How do we explain it expressing itself as both ever-changing (i.e., made of different transitory moments) and permanent (i.e., made of what persists through all moments and strings them together into one collection)?[8] Consider the associated problem of causation. What is the mechanism whereby a beginning condition becomes bound to a specific, different ending condition by a permanent and universally applicable causal rule?[9]

Turning to physics, how do we explain matter being best described by two pictures: one based on particles (individuals) and the other on waves (collective behavior as a continuous thing in itself)?[10] Given mathematics' remarkable effectiveness in describing empirical phenomena and given also mathematics' basis in set theory, we might expect the cosmos to be composed, at least in part, of collections made up of individuals that are themselves collections.[11] But this begs the question of where this set-theoretical behavior comes from and, moreover, puts the paradoxical foundation of set theory squarely at the center of mathematical models of the universe.

Recall also our philosophical problems understanding the basis for empirical science—our inability to adequately explain how here-and-now observational data (which are individual in nature) can give us public theories or models of reality applicable to more than just one here-and-now moment.[12]

This publicity-presencing process becomes especially perplexing and troublesome when applied to ourselves. I, and presumably others, perceive a world around myself made of properties that are public; they're there for others as well as myself. But like Descartes, I'm also entirely convinced on logical grounds that my mind creates for me these properties that I perceive. But how, then, can these properties that I perceive truly be public? How can this desk, this clock, this lamp, this pen I see before me be creations of my mind yet be there for others to perceive as well? How, at the *same* time and with respect to the *same* things, can my faculties of awareness be *both* an individual agent of presence as well as a collective agent of publicity?

This dual agency of my awareness impacts everything I do. Consider something as simple as raising an arm. No doubt I can raise my arm if I want to, but how do I do this?

I can explain the process physiologically in terms of muscle contraction, but this doesn't explain my experience of *me* doing this, how *I* contract my muscles. The act surely has something to do with my faculties of awareness (I can't consciously raise my arm when I'm unconscious). But when I examine my own conscious act of arm raising, I find no explanation for how *I* make my arm go up. I can usually explain *why* I raised my arm (e.g., to stretch muscles or wave at someone), but not *how I* did it.[13]

Again, I find myself with contradictory descriptions of my awareness's activities. On the one hand, "I" am the one who raises my arm, but I'm unable to explain how. On the other hand, my physiology raises my arm, but the explanation of how this happens doesn't capture my own experience of agency. Looked at one way, I find individual agency but no collective agency, and the act seems volitional; looked at another way, I find collective agency but no individual agency, and the act seems deterministic. (You might try this for yourself.)

We've been struggling for a very long time with the various and changing expressions of this riddle of individuals and their collections. It seems fair to conclude that the riddle itself is woven into the fabric of the universe as we know it and into our nature as human beings.

At Home in the Everyday Puzzle of Reality

All the philosophical difficulties mentioned in the last section appear to trace to a single individual-collective puzzle that I've associated with the nature of awareness.

As a puzzle associated with awareness, it concerns each of us *intimately*; it applies to our own moment-by-moment experience of *everything*. The same awareness that presents us with public properties seems also to establish us as distinct individuals—creative agents in our own right. With respect to the same things and at the same time, we seem to be *both* conscripted agents for what is so for all *and* self-willed agents for what is so for just ourselves.

The puzzle may be intellectually troublesome, but we seem to have no difficulty melding the individual and the collective in our every conscious act. As I've suggested in my working definition for awareness, these dual activities seem basic to awareness's nature. We seem quite at home with awareness's two roles in our everyday lives. For me to see what is before me as a desk or hear what is outside as a bird is for me to be presented in this here and now with *one* individual instance of *all* desks or *one* individual instance of *all* birds. The relationship between particulars and universals may have troubled us for a very long time, but it's a relationship that my awareness spontaneously and effortlessly establishes in its every perceptual act.

Likewise with the choices we make—my choice to raise an arm, for example. Instead of raising my arm, I might have just thought about doing it or moved my foot or done something else. But out of all such possibilities, I picked this particular act and made its properties present, its properties the ones instantiated here and now. A choice binds an individual moment to certain collective properties and makes what is present an instantiation of them.

We also merge the individual and the collective in speech and writing. Each word I'm using now—"word," "using," "now"—is a collective construct made concrete and particular in the written context.[14] Likewise with our inventions, our works of art, our crafts. In this individual here and now, we make things that are public—that are there for others as well as ourselves.

The effortless way awareness unites the individual and the collective in our everyday lives brings me to a question. Why should we *care* about the logical problems it causes?[15]

We care because awareness comes not just with the ability to presence publicity; it also comes with an *interest* both in presencing publicity and in what is thus presenced. Our faculties of awareness act *intentionally* to discriminate among things, assign value to what is thus discriminated, and respond according to the value assigned. We recognize what we see as "ice cream," we label it pleasurable to eat, and we decide to eat some. We identify a car as heading our way, we see it as dangerous, and we move to avoid it.[16]

With intentions and values, we come to how we feel about things and what prompts us to act as we do. We also come to emotions such as love, hate, joy, sadness, hope, and fear. We come to what Hindus speak of as the inmost covering around Brahman (the "bliss-illusion sheath"), the animating principle of the universe that is the final veil hiding the ultimate basis of everything.[17] Or in Buddhism, we come to the second Noble Truth, which tells us that the driving force of samsara is desire—our intentions and values and the hold they have over our actions.[18]

We also come to fugues in their second, psychological sense and to awareness's individual and collective roles being more than just an intellectual puzzle. When we dispassionately apply our self-reflective human way of being aware to ourselves, it's difficult to find any ultimate purpose or intrinsic value for our lives. As the Judeo-Christian scriptures describe it in the Garden of Eden story, our way of being aware may have given us an exceptional understanding of our world, but it came with a sense of ourselves as mortal beings subject to pain, death, and estrangement from God, the agency responsible for our public world. Hindus and Buddhists, for their part, call this everyday life samsara—the unending round of fundamentally unsatisfying and pointless joys and sorrows. Existence, says the Buddha, is dukkha, or suffering.[19]

We may be at home with awareness's two roles, but not with our customary conception of ourselves as individual selves or souls.

Of course we don't need religions to tell us that everyday life's unending round of joys and sorrows offers no deep or lasting satisfaction, fulfillment, or sense of self-worth; all we need is a little candid self-reflection. The difference with religions is that they also tell us something else about ourselves. In Hinduism and Buddhism, ordinary self-awareness does not reveal the true nature of either ourselves or reality. Abrahamic monotheism may describe our mortal life as real, but our mortal predicament is not all there is to life in this case either.

If these religions are correct, then the person and awareness that we discover in ordinary self-reflection are not final truths of human existence. The question is, How do we discover this to be the case not just as an intellectual exercise but as an ongoing everyday experience?

Gateless Gates

The great path has no gates,
Thousands of roads enter it.
When you pass through this gateless gate
You walk freely between heaven and earth.

From a thirteenth-century introduction to a collection of Zen koans[20]

The pursuit of enlightenment or the presence of God is an unusual quest. Insofar as my analysis is accurate, then everywhere we turn, the path opens before us: every moment of our awareness is a dialogue between the patterns of our public world and our here-and-now first-person views. If we identify with the dialogue, then no gate blocks our way. If we think of ourselves as an enduring version of our first-person views, then every avenue is a dead end.

How do we learn to walk this path between heaven and earth, as the above quote puts it?

Consider in this regard certain spiritual practices. Let's look, first, at those that address our understanding of reality and, second, at those that address the development of virtue.

Understanding of Reality. I've argued that reality as we know it has at its core an intractable puzzle about individuals and their collections. If this is so, then we might expect to see efforts to understand reality aiming toward an experience of the basis for this puzzle.

One very common (and perhaps surprising) practice to consider in this context is prayer. Prayer comes in many varieties, but I'm interested here in just the bare outlines of prayer as an encounter between oneself and an agent of the collective. One party in this interaction is the person praying; the other party is that to which the prayer is directed: a deity, bodhisattva, saint, ancestor spirit, or some other representation of collective agency.

It's easy to disparage such an encounter as imaginary; after all, this meeting surely takes place within the mind of the one praying. But in my context here, this is precisely the point. Prayer acknowledges the collective standpoint within one's own mind and opens up the possibility of one's mind as the place where the individual and the collective converge. Even when we say prayers aloud to some tangible representation of a deity (e.g., a statue, a picture), we still meet the deity within our minds—a deity often thought to know our deepest thoughts and intentions.[21]

Nevertheless, a problem comes with this way of thinking about prayer and, more generally, the worship of deities. Prayer may give us the feeling of our minds as a meeting place between ourselves and collective agency, but insofar as this meeting is structured as one between two separate agencies, it doesn't, by itself, take us much further. Nondual Hindus speak of the version of Brahman that we encounter in this way as *Ishvara*, or our idea of or interpretation of Brahman.[22] In Buddhism, one might pray to a bodhisattva at the beginning of a meditation session, but if the intent of meditation is enlightenment,[23] then meditation proper aims beyond this level. As for theists, such a concept of God will not likely satisfy someone who has studied philosophy or is especially inquisitive about the nature of our universe.

At this juncture, we come to a variety of lesser-known and, to my mind, underappreciated spiritual practices that specifically address the interests of such people. Many of these practices, such as Hinduism's *jñāna* ("knowledge") yoga or Buddhist *vipassanā* ("insight") meditation, teach a detached but mindful mental posture from which one observes how one's own mind works. Much like in this book, practitioners examine, among other things, the nature and origin of their everyday sense of self and the relationship between their faculties of awareness and reality as they know it. These practices, however, are just the first steps. Jñāna yoga aims at an *experience* of oneself as the impersonal, enduring witness in whose mind arises the public world as we collectively know it.[24] Buddhism's vipassanā meditation aims at an *experience* of the selfless, interactive presence in which our public world arises.[25]

We find this same use of the intellect to press beyond the intellect in Zen/Chán Buddhism and its *koan* (Chinese, *gongans*) meditation. This practice uses meditative riddles—*koans*—that can't be answered with intelligence alone. Again, the goal is to "break through conceptual thought" (as one introduction to a collection of koans puts it)[26] in order to attain the experiential basis of the public world presented by our minds.

Theist religions have their own versions of these practices; however, in these religions, the path of understanding poses a special challenge. As discussed in chapter 8, the dualism of Abrahamic orthodoxy offers no built-in resolution to the divide between the individual agency of ourselves and the collective agency of God. Thus Abrahamic spiritual practices emphasizing our understanding typically have less to do with achieving a higher truth about their account of reality than with transcending it altogether.[27] For example, the Christian bishop Gregory of Nyssa (c. 335–395 C.E.) speaks of the path to God as progressing from light (understanding) to darkness (concealment from understanding).[28] It is in the darkness of the "divine night" where one finds God. In Islamic mysticism (touched on in chapter 8[29]), union with God requires *fanā*, an "annihilation" of the psychological self (suggestive, perhaps, of Buddhism's no-self, *anatta*), which, in effect, provides an experience of reality that is at least as much nondual as dual.[30]

Much more could be said about spiritual practices that address our understanding. My point is that all these religions have practices that seek an experience of reality from a perspective that transcends the conventional, a perspective in which the standpoints of the individual and the collective merge and one's awareness comes to identify itself with this individual-collective interaction.[31]

Just as philosophy is not for everyone, these understanding-oriented practices are not for everyone. They are for those with a strong intellectual bent—those driven by curiosity and who see reality as a puzzle to be solved. For those with other interests, there are other practices advocated for attaining enlightenment. (Recall Hinduism's different varieties of yoga.)

Development of Virtue. Central to most religions is the building of community,[32] and part of community building in religion (or COR building in my jargon here) is an emphasis on moral virtue, the religious dos and don'ts. Injunctions most people are familiar with are those against stealing, lying, harming others, committing adultery, and coveting things that do not belong to you. "You shall love your neighbor as yourself" is another important one; compassion for others is a trait that's almost universally promoted in religious doctrine.

It's not difficult to understand how such fostering of virtue directly addresses the gap between the individual and the collective. The virtues advocated by religions are typically those that encourage putting the interests of others and the collective above individual interests.

But there's a catch. As with our efforts to understand reality, virtue poses a puzzle.

Imagine a science fiction scenario in which there's an electronic device that, when grafted to people's head, subtly compels them to act virtuously. It does this by altering their values so that they *want* to act virtuously; the interests of others become more important than their own individual interests.[33]

It's difficult, I think, to call such acts virtuous; they are too much like those of a robot. Virtuous acts seem to require the voluntary choice of aware individuals, which brings us back to the problem of free will and, in my context here, the puzzle of awareness's two roles.

In considering this puzzle of virtue, recall that the issue of free choice doesn't come up as such in Nondual Hinduism or Buddhism.[34] It's a characteristic of dualist worldviews where individual and collective agency are thought of as fundamentally distinct. But also recall that these Asian religions have their own problems in this area. Nondual Hinduism has difficulty explaining maya and why we experience ourselves as mortal beings when we are fundamentally Brahman. Buddhism has difficulty explaining why a reality without selves or permanence wouldn't also be without karma or spiritual growth or even enlightenment.

Also consider what virtue actually asks of us. Consider, for example, compassion—how genuine compassion for others doesn't seem possible without a *distinction* between ourselves and others. How can we love another across the boundaries that divide us unless there are boundaries that divide us? Virtue seems to require us to be fully individual *yet also* able to understand the world from other points of view and able to feel, at least to some extent, what others feel. Virtue does not seem possible unless awareness can hold the individual and collective standpoints as *distinct* yet *also unite* them in our feelings, choices, and interactions with others.

Moving On

If it's true that accounts of reality conceal as well as reveal, then what of my own account here? What of the mystery this book's point of view conceals?

Perhaps this is an unfair question to ask of me; how can I know what my individual point of view conceals?

Then again, neither my point of view nor my understanding is static. This book has gone through many revisions, and this question might prod the process further. I could explore, for instance, what might be hiding behind the ambiguities of my terminology: a concept of "presence" that's both effable and ineffable at the same time and a concept of "awareness" whereby awareness simultaneously and with intention generates presence, publicity, their union, and their difference. I could look among other accounts of reality for new themes to add to reality's fugue (Chinese philosophy is very promising in this regard[35]). There are also important topics I've not dealt

LIFE, LAUGHTER, AND THE MARGINS OF REALITY

Why do we laugh? What does humor tell us about ourselves?

Foremost among the several historical answers is the "incongruity theory." What we find funny is what "violates our standard mental patterns and normal expectations."[36] Most of our everyday experiences have an orderliness to them. Even surprises usually make sense to us; they fall within the normal order of our world. The incongruity theory argues that humor has its source in surprises that appear to fall outside the normal order, when reality becomes bizarre, often by our own invention. If reality is like a container in which we put all that is fixed and sure in our world, then humor is to be found at the rim of this container. Here on the rim dance Harlequin, Pierrot, and Columbine, turning social norms upside down, just as the culturally subversive Dionysus did well before them. Here are Zen koans undoing, often with a laugh, our minds' conceptual lockstep.

In our own lives today, we find this humorous incongruity in the creative absurdity of political humor, where public figures are characterized in ways that are both quirky and apt. We find it in puns, where the sounds of words are used to make surprising yet somehow fitting associations among word meanings. No doubt incongruity isn't sufficient to explain what makes us laugh; incongruities can also be seen as threatening or simply absurd. Nevertheless, it does appear to be a good start, especially when coupled with the idea of "play."[37]

I've proposed that awareness in its many forms generates our universe's set-theoretical behavior, its logic. In this regard, consider what this incongruity theory implies. We *enjoy* certain violations of this logic—breakdowns in "our standard mental patterns and normal expectations." Something about the discovery of these contradictions in our everyday world and our role in generating them *pleases* us—perhaps in part because they give us a taste of a larger reality that encompasses logic yet moves beyond it as well, a larger reality that appears to be a happier place to live than the one of logic alone.[38]

with. What, for example, might humor and laughter teach us about reality, values, and enlightenment? Even in great suffering, humor (if one can find it) seems to ease the difficulty by turning the circumstances into some variety of play by finding therein something "incongruous" (again, a sort of *puzzle*, as the "incongruity theory" of humor argues; see text box).

Nevertheless, if you find yourself, as I do, face-to-face with a reality that cannot be what it is without it having a certain mystery at its core, then I doubt that another revision of this book would show you your way forward any more clearly. I might want to revise the book to make it clearer to others, but at this stage of my inquiry, much of what this book conceals comes simply from it being a book. Where in these pages do we find God or Brahman or nirvana except emptied of life, shrunk down to fit comfortably into concepts? Where are the sorrows and joys of a true spiritual or philosophical journey? To very loosely paraphrase William James, a book like this might offer perhaps an interesting recipe, but not a real meal.[39] It gives us words but not their meanings, not what we might aspire to on the other side of the gateless gates.[40]

It's rather in the living of our lives where we might come to that place from which philosophies, religions, and accounts of reality emerge and we might discover what words are talking about. It's in our interactions that our world and everything in it come to have meaning and value—some from ourselves, some from others, but mostly from the spaces between us and the relationships among ourselves that we build into these spaces.

In these spaces between us, human biology becomes friends, companions, lovers, spouses, workmates. Stone and wood become buildings, then neighborhoods—houses, then homes. Here between moments of time, sounds become music, words become poetry, movement becomes dance. Here, too, stars and planets become, if not gods, then at least emissaries of light from a larger cosmos that is, in some sense, also us.

In these spaces between us we also discover love and wisdom, humor, laughter, and play. Suffering, or dukkha, remains but becomes something to be transformed by our responses to suffering and our work in the world. So, too, with our words and symbols: the Sacrifice, the Temple, the Cross, the Ka`ba, the Ankh, the Om, and the Wheel of Life become forces of nature opening windows into another realm, not by themselves, but in our individual and collective awareness of them.

It's also in these spaces between us that I think we find the heart of religions—or at least what they have at times been for us. It's in spiritual practice where we come to the mysterious place where the boundaries between ourselves and others both dissolve and do not dissolve at the same time and where the promises made by religions about happiness, wisdom, life meaning, and destiny might be fulfilled.

Postscripts

Postscript One
Scale as a Dimension of Reality

If awareness expresses itself in space and time as occasions of awareness joined together into collections that can themselves be members of other collections, then we don't live in just the four dimensions of space and time. We also live in a dimension of scale. To locate an occasion of awareness in our universe requires not just coordinates of space and time but a coordinate of scale as well.

Scale conceived of as a fundamental dimension of the universe has somewhat the same meaning as when it's used to indicate different sizes of the same thing—the size of a model car, for example, in relation to the full-size car or the size of a lake on a map in relation to the lake's actual size. In my context here, however, scale pertains to the publicity-presencing process of awareness, which implies some differences in how the term is used.

To understand these differences, recall my phenomenological examination of reality and our first-person experience of space and time. Our first-person awareness of the world we perceive around us comes with a sense of our being here and now, and we presume this to be the case for anyone with a first-person view such as ours. However, this experience of being here and now is not just one of space and time; it's also one of scale. I'm aware at my scale; a hummingbird or elephant is aware at its scale. And this location at a size as well as at a here and now appears to be an essential aspect of what it means for a human or any creature to be aware—to presence publicity.

Consider next our third-person space, time, and scale—space, time, and scale as they are for us collectively. In this same-for-all sense, they together might be thought of as something like a container or matrix within which individual instances of here-and-now-and-size are plotted and play out. We don't know most other awareness locations except as possibilities. Our collective universe is that which encompasses all these space, time, and scale options, including those of matter. This is Newtonian space and time modified by Einstein into the single concept of space-time, to which, it seems to me, we must add a dimension of scale if we are to accommodate locations and distances of scale as well as of time and space.[1]

If, in fact, scale is a fundamental dimension of our universe and space and time are consequences of a publicity-presencing process occurring at different scales, then how might differences of scale impact measurements of space and time?

Imagine a group of researchers in a "scaleovator," a device that traveled through scale rather than through space or time—something like in the movie *Fantastic Voyage*.[2] Perhaps a more modern version of their vehicle might be a capsule that enclosed around its occupants and took over their sense organs to create a virtual experience of expanding or shrinking in size. As these scale travelers expanded in size to that of, say, an elephant or whale or beyond to that of a planet or galaxy, objects outside would appear to become smaller, and the distances between them would appear to shrink. As the travelers shrank in size to that of a bird, insect, bacterium, or atom, the reverse would happen: objects outside would become larger and distances would increase.

Since this is scale travel in a universe of space-time rather than just space, imagine that the capsule also altered the travelers' sense of time in the same proportion as their sense of spatial distance. Because time measurements would expand or contract with the changing size, the scale travelers would measure the velocity of a steadily moving object to be the same regardless of their scale. An object moving one centimeter per second at one observational scale would appear to be moving one centimeter per second at one-quarter the observational scale (although the object would also appear to be four times bigger in each dimension and going four times farther in four times the length of time).

Notice how this measurement of velocity differs from comparable measurements in everyday life, where scale differences don't involve a dimension of time. For example, if we measure the velocity of a steadily moving object across the viewing screen of a digital microscope or telescope, we will find its velocity magnified along with the object's size. An object viewed through a 4x lens will move across the screen four times faster than it would without magnification.

These calculations highlight how space and time measurements (and, by extension, mass and/or force measurements[3]) would differ for observers at different scales. Where, then, might we find evidence for such a relationship among scale, space, and time in what we know about living beings and matter?

In fact, we've discovered that many biological aspects of living organisms vary with their size. We know, for example, that size affects the pace of life; consequently, if we want to decipher animal communication, we can typically do so more easily if we slow it down or speed it up depending on whether the creature is smaller or larger than us.[4] We've also learned that the properties that vary with size generally appear to scale in four dimensions rather than three. In 2001, the journal *Nature* gave the following overview of biological scaling laws:

> Naturalists have long known that many aspects of organisms' biology vary with their size. Bigger animals live at a slower pace. They survive longer, grow more slowly, have slower heart rates, and so on. In the 1930s, physiologist Max Kleiber of the University of California, Davis, put a number on this trend. He showed that an animal's metabolic rate is proportional to its body mass raised to the power of 3/4. This relationship has been found to hold across the living world from bacteria to blue whales and giant redwoods, over more than 20 orders of magnitude difference in size. Scaling laws based on exponents in which the denominator is a multiple of four apply to a host of other biological variables, such as life-span. "There's maybe 200 scaling laws that have quarter powers in them," says [Geoffrey West of Los Alamos National Laboratory, New Mexico].[5]

The 4 as the denominator in the exponent of 3/4 tells us that scale affects values in four dimensions. Given our everyday way of scaling mass as proportional to spatial volume only, it's been difficult to explain this exponent—to explain why, for example, metabolism shouldn't be proportional to mass to the 2/3 power instead of to the 3/4 power. These scaling laws have been controversial not only because of the 4 in the denominator but also because they don't apply precisely or in all cases. However, variations like these are also what we'd expect if they were a function of a creature's publicity-presencing process and not strictly of its size.[6]

In any case, no one disagrees that, generally speaking, the larger the organism, the longer the life, as well as the slower the pace of life, heart rate, and growth rate.

Biological scaling laws give us an idea of how scale impacts living creatures. Where might differences in scale play a role in physics?

Before answering this question, I want to emphasize that scale as a fundamental dimension of the universe pertains to the *origin of publicity* in our universe (i.e., to CORs and their members). It doesn't pertain to the size of the things of which we're aware unless they are COR members. Physical objects that come into view at scales much larger than our own—planets, stars, galaxies, and so on—are characterized by properties that seem mostly if not entirely traceable to regularities that apply equally at much smaller scales.[7] Perhaps the way galaxies are arranged in a bubble-like pattern, or the way the universe is expanding, or the way the Higgs field seems to permeate the entire universe at a nonzero base value might turn out to have a larger-scale origin (at least in part). But at present, I'm not aware of any evidence pointing to such an origin for these phenomena.

We do, however, find a concept of scale at the center of thermodynamics and our understanding of energy and work. Thermodynamics's concept of entropy is defined by Ludwig Boltzmann in a way that entails a distinction of scale. He defines the entropy of a system in terms of the number of possible ways its small-scale components can be arranged without changing the large-scale system itself (i.e., the number of "microstates" that will produce a certain "macrostate").[8] Thus, for example, the entropy of a glass of water containing ice cubes is less than that of the glass of water after the ice cubes melt because there are far fewer possible arrangements of water molecules when the water is divided into solid and liquid portions than there are when the water is undivided and mixed together.

Note that entropy's distinction between microstates and macrostates is a distinction in the scale of two frames of reference, two observers. The question I've asked in this book is how we might define such frames of reference.[9] As an answer, I've proposed that they be defined in terms of how and where publicity is presenced, which makes the distinction between microstates and macrostates a distinction between the scale of two CORs. (Recall weak emergence in this regard, which I described as pertaining to realities that arise from the interactions of members of existing CORs at different scales.)

Apart from entropy, it strikes me as at least suggestive of a dimension of scale that one of the principal problems in physics today directly involves scale—namely, our difficulty reconciling general relativity's account of large-scale phenomena with quantum mechanics's account of small-scale phenomena.[10] Also suggestive is string theory's claim that this problem might be solved if additional spatial dimensions existed that came into play at small scales but were concealed from us at our scale.

Also of note is the development in mathematics of fractal geometry, which is devoted specifically to scale. While I'm not aware of any direct evidence as yet from fractal geometry for a natural dimension of scale in the sense I've proposed, its geometric dimension of scale has already found many applications in our everyday world.[11]

Postscript Two
A Definition for Truth

There are many theories of truth, but the most venerable by far is the "correspondence" theory. This theory, which goes back at least as far as Aristotle,[1] also seems to fit best with what we typically mean by the word "true" in our everyday lives. It defines truth as what accords with fact or reality; a statement or idea is true if it expresses what is in fact the case. The statement "It is snowing in New York" is true if it is indeed snowing in New York; otherwise, it is false.

In spite of its pedigree and intuitive appeal, the correspondence theory has had an increasing number of critics over the previous century, and there are today many rival theories. The chief difficulty has been specifying exactly what is meant by this "correspondence"—what is it that a statement is supposed to correspond to, and what is the nature of this correspondence? For example, the statement "It is *true* that it is snowing in New York" doesn't seem to say anything more than the statement "It is snowing in New York."[2]

One of the benefits of defining reality in terms of publicity and presence is that it suggests a definition for truth that should help with understanding not only the correspondence theory but other theories as well. The definition has two senses for truth:

> TRUTH: (1) the publicity of reality considered apart from presence, or (2) a correspondence to that publicity.

Notice that I define truth not as a correspondence to reality but rather as a correspondence to the *publicity* of reality. This setting aside of presence in the definition intends to reflect our everyday use of the word "true" to refer to the *properties* of reality (i.e., publicities)—what can be said or thought *about* what is real—as opposed to reality itself.[3]

To illustrate, compare two questions, one related to reality, the other related to truth. If I asked you if you thought UFOs were *real*, I expect the question would make sense to you. It asks if you think UFOs in fact exist, if something fitting the definition of "UFO" has a presence in our universe and could, at least in principle, impact us or others if there were contact. On the other hand, if I asked you if you thought UFOs were *true*, I expect you might find the meaning of "true" unclear here and the question at least a little odd. To ask what is true in a more meaningful way, I would instead ask *about* UFOs, about their properties (i.e., publicities). For example, I could ask you if it were true that a UFO once landed in your backyard. In other words, is it true that the property "landed in your backyard" holds for some real UFO? I could also ask a question very similar to the one about whether UFOs were real by using "existence" as a property; I could ask if you thought it was true that UFOs existed.

This distinction between real and true applies in other contexts. If I wanted to know if Socrates was an actual historical person, I would ask if Socrates were "real," not if he were "true." On the other hand, if I wanted to know what properties characterized the real Socrates, I might ask what is true or false *about* this reality. "Is it true that he lived in Greece? Died at age seventy? Tutored Plato?"

The same is the case with more abstract realities. I could ask, for example, if the game Monopoly is played with real money (a question that seems to make sense even though both "real money" and "play money" are generated by the human mind[4]). Note again that asking if Monopoly is played with "true" money seems odd. However, I could ask what is true *about* Monopoly money or real money—what it looks like or what its denominations are. Again, questions of truth are about properties, the publicities that characterize realities.

Thus in everyday usage, questions of truth appear to be different from questions of reality. If we frame this difference in terms of publicity and presence and this book's definition of reality, then we can say that questions of reality ask what is both public and capable of presence, while questions of truth ask what publicities characterize these realities.[5]

Publicities not only make up one of reality's two aspects; they are also things that an individual or group may be wrong about. A judgment about the properties of a reality may not match up with the properties as produced by the collectivities of reference. Because we can make false judgments about reality, "truth" concerns not just the public nature of realities (truths in the first sense) but also our ability to discern that nature. Truth in this second sense is the degree to which an individual's or group's judgment of publicity *corresponds* to the publicity as generated by the collectivities of reference. If I'm traveling to New York City today, I might like to know if it's snowing there—if the predicate "is snowing" corresponds to the physical circumstances I would encounter were I in New York City.[6]

Note that judgments of truth include getting the collectivities of reference correct. Simply believing that the earth is flat and at the center of the universe does not make it true regardless of how many share the belief. In this case, a judgment of truth concerns physical realities, not human realities, and it's in the context of material CORs that this issue is decided.[7]

In sum, truths in the first sense refer to the publicities that make up realities (including their CORs), while truths in the second sense refer to correct judgments about truths in the first sense (and correct judgments about their CORs). On the one hand are the properties—the truths—of our world about which we talk and to which words refer. On the other hand is the accuracy—the truth—of our beliefs about these properties and their CORs.

A nuanced analysis of how this theory of truth relates to other theories is outside the scope of this postscript, but I would like to offer a number of observations that might serve such a project.[8]

First, because truth as defined here refers to a reality's publicities and because the presence aspect of a reality is at least partially ineffable, there is in most cases little difference between a statement about the properties (i.e., publicities) of a thing (e.g., the cup is broken) and a statement about the truth of these properties (e.g., it's true that the cup is broken). This characteristic of reality is reflected in "deflationary" theories of truth. As those advocating these theories point out and as I mentioned earlier, adding "is true" to an assertion like "It is snowing in New York" seems to do little more than reassert the original statement. To assert the truth of a statement is to affirm that the statement characterizes the reality it purports to characterize—that snowy weather is present for people in New York and would also be for us if we were there.

Second, because truths are publicities and publicities behave in set-theoretical ways, a number of things follow. As the "coherence" theorists point out, being true implies being consistent with other truths. In other words, a truth is (or is intended to be) part of a web of truths that must remain logically consistent (in at least some sense) if they are to remain invariable over collections of aware individuals. Another consequence of publicities' set-theoretical behavior is that truth can also refer to the logic itself of publicity. Truth in this respect is developed by more formal, mathematical theories, such as the truths given by Boolean truth tables.

Third, our use of words maintained by one COR (the community of discourse, or COD) to refer to realities generated by other CORs implies that, at least in some cases, truth can be analyzed in terms of language's capability for signification. Alfred Tarski's "semantic" theory, for example, examines the semantics of thought and speech insofar as simple sentences in one language (the "object" language) might be referred to as true or false by statements in a second language (the "meta" language).

Fourth, many of our most important truths are publicities generated by our own human collectivities of reference. In other words, they depend on such things as human language, culture, and history. This point is emphasized by "constructivist" and "postmodern" theories. As they point out, truth can in fact be nothing more than what we say it is, or what those in power say it is, or what we agree it is. (Again, in my context here, it's important to keep CORs straight. An asserted truth is "truly true," so to speak, only if it doesn't implicate other CORs that would invalidate it. Truths of chemistry, for example, are ultimately determined by matter, not people.)

Fifth, because publicities are what remains the same for collections of awareness occasions, we discover publicities—a.k.a. truths—of the universe we live in by discovering those aspects of our universe that remain invariable for us as human beings. This point is made by "pragmatic" theories, which argue that truth is what "works," or is empirically verified, or will eventually turn out to be the case once all the evidence is in. If a theory of physics predicts that gravity will bend light, then that theory is "true" for us insofar as our experiments verify that light is affected by gravity in this way. (Note the two CORs involved in this example. The truth we seek is what is true of *matter* insofar as it might be known by *human beings*.)

Sixth, because truths are about participation in collectivities of reference, they very much concern our personal agreements with statements about reality, whether one opts in or out of various collectivities of reference. To say "It is true that it is snowing in New York" is to say, in part, "Yes, I agree with those for whom that statement describes reality." This point is emphasized in "performance" and "consensus" theories of truth.

Seventh, truth as publicity suggests not one but several kinds of truth that depend on what publicities are of interest. On the one hand, an interest in truth can reflect our everyday efforts to ascribe the right properties to what we perceive around us. "Is it true that the river is polluted?" "Is it true that we're supposed to get two feet of snow tonight?" A simple question such as "What is that over there?" is likewise an interest in truth, an interest in what properties characterize the reality you are pointing out to your companions.

On the other hand, philosophers often mean other things by truth. The notion of truth can express an interest in the publicities that hold for the widest collective scope. In this guise, truth can reflect the advance of science and philosophy as projects of finding truths that are progressively more universal and are not limited to certain groups of people or certain forms of awareness. Examples would be physicists' efforts to find a "theory of everything" or my own effort here to point to an underlying mystery that might help explain certain differences we find among philosophical, religious, and scientific accounts of reality.

Truth in this philosophical sense can also express our interest in "truth" itself, what the word "truth" means, what truth *is*. This is truth, for example, in the sense that I have been discussing in this postscript.

Yet again, especially in religions, it seems to me that truth in this philosophical sense can refer to what lies beyond any discursive account of it, what one understands only through actual experience. This sense of truth refers not so much to some knowledge or understanding to be realized as it does to the realization that grants this understanding. If truth is publicity and publicity lies in the network of relationships that make up reality as we know it, then the practice of living out of the truth as best we know it is no different from the practice of living in the spaces between us where awareness constructs our universe.

Acknowledgments

A number of people have helped me develop my ideas for this book and communicate them simply and clearly. Randall Davis and Lee Goldberg met with me almost every week for six years to listen to me read chapter rewrites. My text in that early period was often abstruse and taxing to listen to; yet, to the end, they remained willing participants and made substantial contributions to the first draft.

I put the first draft to the test in an adult evening school course, whose class members demonstrated that there was an enthusiastic audience for the book's subject matter. The text, however, did not satisfactorily explain my approach, and the students encouraged me to persevere in a rewrite and offered many helpful suggestions. Two class members volunteered to help me with the second draft.

My partners through the long rewrite were Carol Parker, Francis Riesz, and Fran Perlman. Carol Parker brought to the project professional editing talents and a remarkable intuitive understanding for both philosophy and the perspective on reality I was proposing. Francis Riesz contributed his academic background in philosophy and his skepticism; he pointed out gaps in my arguments and unwarranted assumptions and sought to keep my ideas reasonable within a larger philosophical context. Fran Perlman added a rare combination of philosophical insight and common sense, as well as an exceptional knack for clarifying and polishing my wording. Along with Carol Parker, she made certain that the text was understandable to someone interested in philosophy but who lacked academic training in that field. In addition, Dwight Harris's philosophical background as well as his continuing interest, comments, and encouragement throughout the writing of the second draft have been very much appreciated. Special thanks also goes to Peter Meyers for his careful reading of the book and his insightful, helpful criticisms and comments not just in the sections on physics but throughout. Others who contributed their expertise in various respects were Philip Clayton, Stephanie Theodorou, William Grassie,

Wendy Middleton, Kathryn Yahner and her colleagues at Penn State University Press, Max Crandall of CranCentral Graphics, and Scribe Inc.

Much of the credit for this project also goes to Susan Bindig, who provided much more than her opinions and editing skills. Throughout the long and challenging writing process, she has given me companionship, a different perspective on life to balance my own, and the space in our joint lives for my philosophical pursuits.

SAMUEL BRAINARD

Terms Defined in This Book

Principal Terms

AWARENESS: The presencing of publicity. Awareness (1) establishes presence (its disjunctive role), (2) cooriginates/comaintains publicities (its conjunctive role), and (3) joins publicity with presence to produce realities (both roles together). All three elements are interrelated; anything that acts in one of these ways is aware and acts in all these ways. (See chapter 10 section "A Definition for Awareness.")

PRESENCE: The product of awareness's disjunctive role. The simple "thereness" of whatever we perceive; the arising of properties in awareness irrespective of what the properties are (what Buddhists call *tathatā*, or "suchness" or "thusness"). Presence is defined from the standpoint of the individual and the first-person view; it expresses itself as pure particularity as well as here-and-now interaction in space, time, and scale. Presence attempts to name something that is at least partially ineffable; it signifies what would remain of a reality were we conceptually able to remove everything signifiable about it (all publicities). (See chapter 6 section "Enlightenment and 'Presence'"; chapter 7 section "Reality as What Is 'Present'"; and chapter 9 discussion of presence in section "Sitting in a Café.")

PUBLICITY: The product of awareness's conjunctive role. Anything originated/maintained as the same from one occasion of awareness to another. Publicity is defined from the standpoint of the collective and the third-person view; it refers to what is invariable for members of a collectivity of reference (COR) collectively irrespective of what might be present for members individually. Publicities encompass universals (e.g., words, word meanings, the regularities of physics) as well as particulars (e.g., the clock on my desk, Socrates). (See chapter 5 section "A Phenomenological Illustration of First-Person Philosophy"; chapter 6 section "Universals as Arising through Causal Interactions"; and chapter 7 sections "Reality as What Is Public" and "Individual and Collective Agency.")

REALITY: Publicity conjoined with the capability for presence. Realities originate with awareness of various kinds. They encompass publicities with here-and-now presence in awareness as well as publicities with the potential for presence under different circumstances (which casts presence as a property of reality that in most cases must be figured out in conjunction with a reality's other properties). (See chapter 7.)

Other Terms

AWARE INDIVIDUAL ("SELF" OR "SOUL"): Awareness identifying itself as disjunctive in the dimension of space (the individual here) and conjunctive in the dimension of time (a collection of prior and anticipated occasions of awareness, an enduring being). Our everyday experience of ourselves as opposed to who/what we might ultimately be. (See chapter 9 section "Who Are We Really?" and chapter 10 section "Our Concept of 'Self' in Space and Time.")

CLASS: The collection of a publicity's instances, as distinguished from a COR, which gives an origin of that publicity. Anthills, for example, distinguish a class of objects built by, and thus having a COR of, ants. (See chapter 10 section "A Mosaic of Groups: CORs, Classes, and CODs.")

COLLECTIVITY OF REFERENCE (COR): A collection of aware individuals or occasions of awareness that originate and maintain realities. COR members can be separated in space and/or time (e.g., spatially separated people, temporally separated awareness events). CORs themselves, along with the realities they generate, can also be separated in scale. See *scale*. (See chapter 10 section "A Mosaic of Groups: CORs, Classes, and CODs.")

COMMUNITY OF DISCOURSE (COD): A COR that originates a reality (a signifier) that is used to indicate another reality (the signified) usually maintained at least in part by other CORs. (See chapter 10 section "A Mosaic of Groups: CORs, Classes, and CODs.")

CONJUNCTIVE AWARENESS: The role of awareness that originates publicities. In concert with other occasions of awareness, it groups together different here and nows according to what they hold in common. For humans, the activities of this role are largely unconscious and involuntary. This role coupled with awareness's "disjunctive" role generates realities. (See chapter 9 section "Awareness's Two Roles.")

CONSCIOUSNESS: The combined activity of awareness's two roles in human beings, distinguished, at least in part, by human awareness's capacity to use language to understand realities that would otherwise be outside the scope of its presencing capabilities. (See chapter 5 section "Awareness and Conscious Awareness"; chapter 9 section "The Consciousness Problem"; and "The Term '*Chlamydomonas reinhardtii*'" in the chapter 10 section "A Mosaic of Groups: CORs, Classes, and CODs." For the Asian context, compare the concept of *vijñāna* in Nondual Hinduism [chapter 4 section "Brahman"] with this concept in Buddhism [chapter 6 section "The Nature of Reality"].)

DISJUNCTIVE AWARENESS: The role of awareness that generates presence. It provides a here-and-now, first-person arena for experiencing and interacting. For humans, this role expresses itself as conscious awareness's distinctive "attention" or "focus" and as the experience of choosing and acting voluntarily. This role, coupled with awareness's "conjunctive" role, generates realities. (See chapter 9 section "Awareness's Two Roles.")

SCALE: A dimension besides those of space and time in which CORs and the realities they generate are differentiated by the relative spatiotemporal "size" of their COR members. "Size" is not just a spatial volume but a space-*time* volume related to awareness's publicity-presencing process. (See chapter 11 section "Reality as Emergent" and postscript 1, "Scale as a Dimension of Reality.")

TRUTH: The publicity of a reality considered apart from presence, or a correspondence to that publicity. (See postscript 2, "A Definition for Truth.")

Glossary of Hindu and Buddhist Terms

All Indian philosophical schools share Sanskrit concepts. The "Hinduism" and "Buddhism" designations below indicate where the terms play central roles.

ADVAITA-VEDĀNTA (ad-VUHEE-tuh vay-DAHN-tuh). *See* Nondual Hinduism.

AGNI/AGNI (UHG-nee). Hinduism. Sanskrit for "fire," especially the ritual fire, as well as for the fire deity.

AMITĀBHA/AMIDA BUDDHA (ah-MEE-tah-ba / ah-MEE-dah; *Amitābha* in China and Tibet, *Amida* in Japan). Mahayana Buddhism. Popular bodhisattva who promises rebirth in his heavenly Western Paradise (*Sukhāvatī*) for all who have sufficient faith in his power and compassion.

ĀNANDA (ah-NUN-duh, Sanskrit). Hinduism. Absolute bliss. A quality identified with Brahman.

ANĀTMAN (uhn-AHT-muhn, Sanskrit). See *anatta*.

ANATTA (uhn-UH-tah, Pāli; Sanskrit: *Anātman*, uhn-AHT-muhn). Buddhism. "No-self." One of the three marks of existence, along with *anicca* (impermanence) and *dukkha* (suffering).

ANICCA (uh-NIK-uh, Pāli; Sanskrit: *anitya*). Buddhism. "Impermanence." One of the three marks of existence, along with *anatta* (no-self) and *dukkha* (suffering).

ĀTMAN/ĀTMAN (AHT-muhn). Hinduism. The underlying true self. In Nondual Hinduism, considered to be identical to Brahman and, when used in this sense, often capitalized.

BHAGAVAD-GĪTĀ (BUH-gah-vahd GEE-tuh). Hinduism. A short section in the *Mahābharata*. Hinduism's most respected and popular scriptural text.

BHAKTI MARGA (BUHK-tee MAR-gha). Hinduism. The path of bhakti yoga practices.

BHAKTI YOGA. Hinduism. The yoga of love and devotion.

BODHISATTVA (bohd-hee-SAHT-vuh). In Mahayana Buddhism, one who is very close to enlightenment but postpones nirvana for the sake of helping others achieve enlightenment.

BRAHMA (*Brāhma*, BRAH-muh, Sanskrit). Hinduism. The Hindu god of creation. One of the three deities in the threefold (*trimūrti*) of Brahma-Vishnu-Shiva.

BRAHMAN (BRUH-muhn). Hinduism. God as the fundamental nature of the universe. Also *sat-chit-ānanda*, or being-consciousness-bliss, as well as the nameless, or what cannot be named or described by any properties.

BUDDHA (BUHD-uh). Buddhism. "Enlightened one." A title for Siddhartha Gautama.

CHANDRAKĪRTI (CHUN-druh-KEER-tee). Mahayana Buddhism. Prominent figure in the Mādhyamika school.

CHIT (*cit*, CHIT, Sanskrit). Hinduism. Universal or archetypal consciousness, awareness, or knowledge. That by which we know what we know. A quality identified with Brahman.

DHAMMA/DHAMMA AND DHAMMAS (DUH-muh, Pāli). See *dharma*. In Theravada Buddhism, dhammas are the psychophysical force particles that make up existence.

DHARMA/DHARMA (DUHR-muh, Sanskrit; Pāli: *dhamma/Dhamma*, DUH-muh). Hinduism and Buddhism. In Hinduism, right action; ethical duty; that which determines righteousness and morality; the basis of all religions. In Mīmāṃsā Hinduism, the inherent, primordial potency of ethical action. In Buddhism, the teaching of the Buddha; the path to enlightenment; cosmic law. See also *dhammas*.

DUAL HINDUISM (*Dvaita-Vedānta*). The Hindu Vedānta school that claims that Brahman differs fundamentally from the ātman of each individual. Brahman in this school is Ishvara, or the supreme personal God, which is much like the conception of God in Western monotheism.

DUKKHA (DOOK-kah, Pāli; Sanskrit: *duḥkha*). Buddhism. Suffering. One of the three marks of existence, along with *anicca* (impermanence) and *anatta* (no-self).

DVAITA-VEDĀNTA (duh-VUHEE-tuh). *See* Dual Hinduism.

FIVE SHEATHS (Sanskrit: *pancha-kośas*). Vedānta Hinduism. The five layers of maya, or illusion, covering the Ātman. From the outside, these are the (1) "food-illusion sheath" (*annamayakośa*), (2) "life-illusion sheath" (*prānamayakośa*), (3) "mind-illusion sheath" (*manomayakośa*), (4) "intelligence-illusion sheath" (*vijnānamayakośa*), and (5) "bliss-illusion sheath" (*ānandamayakośa*).

IMPERMANENCE (Sanskrit: *anitya*; Pāli: *anicca*). See *anicca*.

INTERDEPENDENT ARISING (Sanskrit: *pratītya-samutpāda*; Pāli: paṭicca-*samuppāda*). Buddhism. The interactive process that gives rise to the everyday world and, with it, what we think of as universals and enduring individuals.

ISHVARA (*Īśvara*, EESH-vur-uh, Sanskrit). Hinduism. "Lord of the universe." Supreme personal God in Hinduism. In Nondual Hinduism, Brahman insofar as understood by conventional reason.

JAINISM. An Indian school of thought not based on the Vedas that denies Brahman and emphasizes the interrelationship among all things. It's prominence as a world religion dates from the Axial period (eighth to third century B.C.E.).

JÑĀNA MARGA (jeh-NAH-nuh). Hinduism. The path of jñāna yoga practices.

JÑĀNA YOGA. Hinduism. The yoga of knowledge and wisdom.

KARMA (KUHR-muh). Hinduism and Buddhism. Moral cause and effect; as we sow, so we reap.

KARMA MARGA. Hinduism. The path of karma yoga practices.

KARMA YOGA. Hinduism. The yoga of work and service.

KARUṆĀ (kah-roo-NA). Buddhism. Compassion.

MĀDHYAMIKA (mahd-YUH-mee-kuh). Buddhism. One of the two main schools of thought that grew out of the Prajñā-pāramitā literature (the other being Yogācāra). It emphasized the emptiness of all accounts of reality in that they were necessarily based on universal first principles.

MĀHĀBHARATA (mah-ha-buh-RUH-tu). Hinduism. The Hindu epic that gives the history of Krishna.

MAHĀSAÑGHIKA (muh-ha-SANG-hee-kuhs). Buddhism. Literally "great/large community." Precursor to Mahayana.

MAHAYANA (muh-ha-YAH-nah; Sanskrit: *Mahāyāna*). Buddhism. Literally "Great Vehicle." One of the two main branches of Buddhism, the other being today's Theravada.

MANAS (MAN-as, Sanskrit; Pāli: mano, MAN-oh). Hinduism and Buddhism. Roughly "mind," but as the faculty by which we receive impressions from the world around us and not as an inner realm of experience separate from the outer world. In Buddhism, the sixth sense organ.

MAYA (māyā, MAH-yah). Hinduism and Buddhism. Illusion; the ignorance that keeps us from enlightenment.

MĪMĀMSĀ (mee-MAHM-sah). Hinduism. A school of thought emphasizing ethical action and priestly rituals.

MOKSHA (*mokṣa*, MOHK-shuh, Sanskrit). Hinduism. "Liberation." Final enlightenment. Liberation from samsara and the eternal round of joy and suffering, birth and death.

NĀGĀRJUNA (nah-GAHR-joo-nuh). Buddhism. Founder of the Mādhyamika school.

NIRVANA (*nirvāṇa*, nir-VAH-nuh, Sanskrit; Pāli: *nibbāna*). In Buddhism, the goal of spiritual practice, which is characterized by clarity of understanding, bliss, and release from the causal cycle of interdependent arising.

NONDUAL HINDUISM (*Advaita-Vedānta*). The Hindu Vedānta school in which the Ātman, the true self, and Brahman are considered identical.

NYĀYA (nuh-YAH-yuh). Hinduism. The school of thought that develops the logical principles used in all Hindu schools. Usually grouped with the Vaiśeṣika school.

PĀLI (PAH-lee, Pāli). Buddhism. The language of the original Buddhist scriptures.

PĀLI CANON (Sanskrit: *Tripitaka*; Pāli: *Tipitaka,* "Three baskets"). Early Buddhist scriptures.

PAÑÑĀ (PUHN-nuh). Buddhism. Wisdom. See *prajñā*.

PRAJÑĀ (PRUHJ-nyah, Sanskrit; Pāli: paññā, PUHN-nuh). Buddhism. Wisdom.

PRAJÑĀ-PĀRAMITĀ SŪTRAS (PRUHJ-nyah-pah-ruh-MEE-tah SOO-truhs). Buddhism. "Perfection of Wisdom scriptures." These writings emphasized the doctrine of śūnyatā, or "emptiness," specifically as applied to all that existed.

PRAKṚTI (PRUH-kri-tee). Hinduism. Matter. One of the two first principles of the Sāṃkhya and Yoga schools, the other being puruṣa.

PURĀṆAS (pou-RAH-nuhs). Hinduism. Popular stories about the gods and goddesses, their battles against demons and interactions with humans.

PURUṢA/PURUṢA (pou-ROO-shah). Hinduism. A "person"; similar in meaning to ātman with a small *a*. In Vedānta, it's equated with Ātman (with capital *Ā*) and thus with Brahman. In the Sāṃkhya and Yoga schools, it is the pristine pure consciousness that is one of their two first principles (the other being prakṛti).

RĀJA YOGA (RAH-juh). Hinduism. Another name for the yoga of the Yoga school.

RĀMĀYAṆA (rah-mah-YUH-nuh). Hinduism. The epic giving the history of Rāma.

ŚĀKYAMUNI (SHAHK-yuh-moo-nee). Buddhism. "Sage of the Śākyas." Another of Siddhartha Gautama's titles; he was a member of the Śākya clan.

SAMĀDHI (suh-MAH-dee, Pāli/Sanskrit). Hinduism and Buddhism. The meditative states close to and sometimes including enlightenment.

SĀṂKHYA (SAHM-kee-uh). Hinduism. One of the two schools (along with Yoga) that conceive of the universe as having two first principles: puruṣa and prakṛti.

SAMSARA (saṃsāra, sum-SAH-ruh). Hinduism and Buddhism. Literally "journeying." The daily round of life, death, joy, and suffering that is our human mortal predicament.

SANGHA (saṅgha, SAHN-gha, Pāli; Sanskrit: saṃgha). Buddhism. Community of Buddhists. One of the Three Jewels.

SAT (SUHT, Sanskrit). Hinduism. Existence, being, that by which we are what we are; a quality identified with Brahman. The related word satya means "truth."

SAT-CHIT-ĀNANDA (saccidānanda, SUHT-CHID-ah-NUN-duh, Sanskrit). Hinduism. Literally "existence-consciousness-bliss." The three qualities identified with Brahman.

SHIVA (Śiva, SHIH-vuh, Sanskrit). Hinduism. The Hindu god of destruction. One of the three deities making up the threefold (trimūrti) of Brahma-Vishnu-Shiva.

SIDDHARTHA GAUTAMA (Siddhārtha Gautama, sid-DHAHR-thuh GAUT-uh-muh, Sanskrit; Pāli: Siddhatta Gotama, sid-HUT-tuh GOUT-uh-muh). Buddhism. Often referred to as simply the Buddha, he is conventionally thought of as the founder of Buddhism, although most Buddhists consider him one in a long line of Buddhas.

ŚĪLA (SHEE-luh, Sanskrit; Pāli: sīla, SEE-luh). Buddhism. Ethical conduct.

ŚŪNYATĀ (OR ŚŪNYA) (shoon-YUH-tuh; SHOON-yah). Buddhism. "Emptiness"; "voidness" of existence.

SURYA/SURYA (sūrya, SOOR-yuh). Hinduism. Sanskrit for "sun" as well as the sun deity.

TAṆHĀ (TAHN-hah, Pāli; Sanskrit: tṛṣṇā). Buddhism. The desire or craving that causes suffering.

TANTRA (TUHN-truh). Hinduism and Buddhism. Schools of thought and associated texts emphasizing esoteric rituals and practices for enlightenment. The practices are very different in the two religions. Hindu Tantra tends to emphasize ritualized behaviors that are taboo in the conventional religion (certain sexual practices; eating meat). Buddhist Tantra involve principally meditation practices involving, for example, manipulating spiritual energy within the body.

TATHATĀ (TUHT-uh-ta). Buddhism. "Thusness"; "suchness."

THERAVADA (Theravāda, tehr-uh-VAH-duh). One of two main branches of Buddhism, the other being Mahayana.

THREE JEWELS OR THREE TREASURES (Pāli: Tiratna; Sanskrit: Triratna). Buddhism. The Buddha, dhamma, and sangha (Pāli/Sanskrit: Buddha, Dharma, and Sangha).

THREE PATHS. Hinduism. The three paths (margas) to liberation (moksha) of karma marga, bhakti marga, and jñāna marga.

UPANISHADS (Upaniṣads, oo-PUH-nee-shuhds; from upa, "supplementary," and ni-ṣad, "to sit down near" a spiritual teacher). Hinduism. That part of Hinduism's scriptures containing philosophical speculations between roughly 800 and 200 B.C.E.

VAIŚEṢIKA (vuhee-SHAY-she-kah). Hinduism. The Hindu school that examines and delineates the general categories humans use to think about and organize experience. Usually grouped with the Nyāya school.

VEDĀNTA (vay-DAHN-tuh). Hinduism. "End of the Vedas." A family of schools that account for the beliefs about ultimate reality held by most Hindus. The three principal schools are (1) *Advaita-Vedānta*, the nondual school; (2) *Dvaita-Vedānta*, the dual school; and (3) *Viśiṣṭādvaita-Vedānta*, the qualified nondual school.

VEDAS (VAY-duhs). "Knowledge." Hinduism's historical scriptures.

VIJÑĀNA (vijh-NAH-nuh, Sanskrit; Pāli: *viññāṇa*). Buddhism and Hinduism. Consciousness; discernment. In Hinduism, the underlying awareness that animates the universe and that enlightenment identifies with our own awareness. In Buddhism, one of the five *aggregates*; it is itself differentiated into six kinds, one for each sense organ (with the mind as a sense organ).

VISHNU (*Viṣṇ*, VISH-noo). Hinduism. The Hindu god of preservation or maintenance. One of the three deities making up the threefold (*trimūrti*) of Brahma-Vishnu-Shiva.

VIŚIṢṬĀDVAITA-VEDĀNTA (VISH-tah-ad-VUHEE-tuh). Hinduism. The "qualified nondual" school of Vedānta, which takes a middle position between the nondual and the dual schools by teaching that Brahman is different from the Ātman in qualities but not fundamentally.

YOGA (YOH-guh; same root as English *yoke*). Hinduism and Buddhism. Literally "yoke." (1) A spiritual practice that aims at constraining or redirecting ("yoking") certain natural tendencies of the mind in order to achieve enlightenment. (2) One of the two schools (along with Sāṃkhya) that conceive of the universe as having two first principles: puruṣa and prakṛti.

YOGĀCĀRA (YOH-gah-CHAHR-uh). Buddhism. One of the two main schools of thought in Mahayana that grew out of the Prajñā-pāramitā literature. The other is Mādhyamika. The signature doctrine is that all phenomena are entirely mental creations arising from within a single consciousness or knowing process. Sometimes called the "mind-only" school.

Notes

Chapter 1

1. For now, I'll use "true for all" as my shorthand definition for reality. As the next chapter will discuss, "real" historically refers to what exists independently of our minds. Used in this historical sense, the term distinguishes naturally occurring things and properties (e.g., physical objects, perhaps the species differences of living creatures) from creations of our minds (e.g., ideas, assumptions, and properties we invent to describe the world). In many cases, however, we don't use the word in this way. Defining reality in terms of mind-independence excludes many things created by the human mind that we today would usually call "real" (e.g., traffic laws, the value of money, music, food recipes).

   Nevertheless, defining reality as "true for all" raises its own questions: What does "true" mean? What "all" does the phrase refer to? Math and logic seem true for all, but are they "real"? "True for all" is thus an approximation, my best for the moment. (Note that by "true for all," I don't mean believed by or known by all. A few centuries ago, many people didn't know that the earth went around the sun, but that didn't keep it from being true for all in the sense intended here.)

2. Freud, *Civilization and Its Discontents*, 11–13, 19, 21–32.

3. I expect this story of rats developing rituals in a Skinner box is true. B. F. Skinner himself did such experiments with pigeons. They were fed irrespective of their own lever-pushing and developed ritualistic behavior shaped by whatever they were doing when they were fed (Skinner, "'Superstition' in the Pigeon").

4. The mind's role in creating reality in Hinduism and Buddhism is discussed in chapters 4 and 6. The rationale behind deities and other "supernatural" potencies is discussed in chapters 3, 7, and 8.

5. This question about whether reality is ever-changing or eternal has been with us at least since Heraclitus's and Parmenides's different conclusions on this issue.

6. Rosenberg, *Philosophy of Science*, 84–91 (esp. 87), 119, 131; Creath, "Logical Empiricism"; Blackburn, *Oxford Dictionary of Philosophy*, 223–224; Goldman, *Science Wars*, 34–36.

7. The question of where to draw the line between science and philosophy may be controversial, but it's hard to deny that there is such a line. The demise of logical positivism in the twentieth century (see "Logical Positivism and the Limits of Science" text box) offers one of the clearest illustrations that empirical science does not explain, among other things, its own philosophical basis.

8. Dianna Eck of the Harvard Divinity School was the first I know to classify responses to religious diversity into "exclusivism," "inclusivism," and "pluralism" (Eck, *Encountering God*, chapter 7).

9. How might skepticism fit into these exclusivist, inclusivist, and pluralist categories? Skeptics respond to our diverse accounts of reality with a thoroughgoing doubt of them all, and such an approach can easily be nihilist if taken no further—a lack of belief in anything. But skepticism comes in other varieties as well. Insofar as this position stems from conflicting accounts that seem equally plausible/implausible, then it fits well in the pluralism category. There are also other versions. It has, for example, been viewed as a precursor to various kinds of enlightenment. The Greek philosopher Pyrrho taught skepticism as a method to achieve peace of mind (Vogt, "Ancient Skepticism," §4). For the Buddhist philosopher Nāgārjuna (to whom I'll return in a later chapter), radical skepticism can lead to *paramārtha satya*, or "higher truth." René Descartes, in his *Meditations on First Philosophy*, describes his own philosophical epiphany as based on a radical doubt (I'll also return to Descartes later). In such cases as these, skepticism would seem to aim at an insight or state of mind beyond conventional categories of thought, including those I've given here.

10. May, *Harvard Dictionary of Music*, 336.

11. I owe the metaphor of a fugue to Douglas R. Hofstadter's *Gödel, Escher, Bach: An Eternal Golden Braid*, which is described on the cover as "a metaphorical fugue on minds and machines in the spirit of Lewis Carroll." In it, we find examples upon examples taken from logic, math, art, and music of how we live in a cosmos where the orderly and comprehensible are repeatedly subverted to our personal enjoyment by contrasting themes. From tortoise-and-hare stories to physics and mathematical conundrums, from Escher's pictures of mind-bending staircases to Bach's musical complexities, Hofstadter shows life to be rationally understandable only as an eternal fugue of pattern and paradox.

12. Why isn't reality-as-fugue an inclusivist account? The approach I've taken here presumes that it is not possible to understand reality fully through one account alone regardless of how inclusive it attempts to be; we've tried that and failed. As an alternative, this book proposes a way of looking at *existing* accounts so that they *themselves* illumine *each other* and together help flesh out a mystery of reality that is outside the scope of any one account.

    Of course, this raises the question of whether or not this "way of looking at existing accounts" is not itself a new account. In this regard, consider Wittgenstein's notion of "language games." This term comes up frequently in arguments against foundationalist philosophy, but what are we to make of the notion of a "language game"? Does not that notion have itself certain paradigmatic qualities? But doesn't it also try to point to another way of thinking about philosophical accounts in general?

13. When we face choices between truth and self-interest, self-interest almost always wins (e.g., Fine, *A Mind of Its Own*, 3–29).

14. The conflict between happiness and the particular kind of knowledge alluded to here is neither a new phenomenon nor one limited to knowledge gained via modern science. Consider the Judeo-Christian scriptures' Garden of Eden story. The primordial couple, Adam and Eve, acquire a certain godlike knowledge, but when they apply that newfound knowledge to themselves, they discover that they are mortal creatures with no more intrinsic worth than the dust of the ground from which they came and to which they will return: "For dust you are and to dust you shall return" says God to Adam (Gen. 3:19).

15. Weinberg, *First Three Minutes*, 154.

16. Religious accounts of reality—theist ones in particular—don't make sense to most Western philosophers. Per a November 2009 survey of philosophers' beliefs about God, only 14.6 percent accept or lean toward theism (Bourget and Chalmers, "What Do Philosophers Believe?" 15).

Chapter 2

1. For "plausible" pictures of reality to feature in this book, I've chosen examples that are widespread throughout the world and have been with us a long time by virtue of what seem to me to be their reasonable presuppositions and their logical consistency within the framework of these presuppositions. Also, as discussed in chapter 1, I presume that science provides humankind's best effort for understanding universal truths, and although science's purview is limited, its findings impact what might be plausible as accounts of reality in not just philosophy but also religion.

2. The concept of "reality" is itself a good example of the inadequacy of our current philosophical vocabulary. As will be discussed shortly, for many philosophers today, the term continues to carry its traditional meaning of "mind independent." However, consider money or political borders or words in a language or food recipes. Surely most of us think of such things as real even though they depend for their existence on the human mind.

3. In the academic community, relative truths are defended by "constructivists," while universal truths are defended by "essentialists" or "perennialists." For an argument that this debate might itself be a symptom of an underlying philosophical puzzle that may in fact be addressed in different ways by various mystical traditions, see Brainard, "Defining 'Mystical Experience.'"

4. The distinction between first-person and third-person views—despite the antiquity and pervasiveness of the grammatical distinction—is fairly new in its current usage as a philosophical distinction. While its exact origins are unclear, it appears to have come into favor with the increased interest in and debates over human consciousness in the twentieth century. Since then, it has largely replaced Descartes's mind-matter distinction in the way philosophers frame the "hard" problem of consciousness (discussed here in chapters 5 and 9).

5. Note that the distinction made here between third-person and first-person views is not the same as the distinction between what a thing is "in itself" and what it is "in relationship to us" that we've inherited from the Enlightenment (as in Locke's distinction between "real" and "nominal" essences of things: *Essay Concerning Human Understanding*, III:iii:15ff., III:iv:9–10; and also Roger Woodhouse's introduction, pp. xvii–xviii). As I use it here, the third-person view includes things in their relationship to us insofar as these relationships are the same for all of us. For example, water characterized as a molecule made of hydrogen and oxygen atoms describes water as a "thing in itself"—water as it appears to exist irrespective of human experience. Water characterized as a clear, naturally occurring, thirst-quenching liquid describes water as it "relates to us"—how humans in general see it, where they find it, and how it satisfies human desires. However, if I want a glass of water to drink, then the third-person view of interest to me is water as it quenches people's thirst, as it relates to human beings and serves human interests. I and other people don't need to know what water is in itself (i.e., H_2O) in order for water to be part of the third-person world we share.

(See also Kant, *Critique of Pure Reason*, B69–B71, B164–B165 [Smith 88–89, 172–173]; and this book's chapter 3 section "Universals, Particulars, and Accounts of Reality." Also see Aristotle, *Metaphysics*, Γ: 6 (1011a3ff.), and in Indian philosophy, King, *Indian Philosophy*, 122–123.)

6. Prior to René Descartes, Greek and Roman philosophers questioned whether our perceptual organs and reason could ever show us reality. The Skeptic school of Greek philosophy, for example, argued that we could not know (or at least not know at the present time) what really existed, and we should refrain from making judgments about the true nature of reality. Objects appear different from different perspectives, and we lack adequate criteria for deciding among appearances. What was not, however, a part of anyone's worldview in the West before Descartes as far as I know was his sharp divide between internal mind and external matter (or between mind and body insofar as our bodies are material in nature). For Greeks and Romans, our perceptual organs were faculties of knowledge that related us directly to the external world, faculties that were (or were not) capable of revealing the true properties of this outside world. For Descartes, what we perceived was not outside of us but within our minds, regardless of how capable our perceptual organs might be (Vogt, "Ancient Skepticism").

7. Rorty, *Philosophy and Mirror of Nature*, 17–61, examines how Descartes and the Enlightenment invented our modern concept of mind.

8. Descartes's *Meditations on First Philosophy* appeared in 1641. Beginning in the prior century, an enormous growth occurred in our Western understanding of human anatomy. Also, at the end of that prior century, around 1590, the microscope was invented and then further developed by Galileo beginning in 1609. Telescopes have been traced to around 1608 and were improved upon also by Galileo beginning in 1609.

9. Descartes, *Meditations on First Philosophy*, 315.

10. There is little doubt that we live our everyday lives as if we directly perceive a reality that is independent of our minds. As a theory of perception, however, this philosophical view ("direct" or "naive" or "commonsense" realism) has few defenders. No one has yet satisfactorily explained how we might perceive a truly mind-independent object without including a mediating process of some kind—a physical conveyance or a cognitive translation between an object and our awareness of it—and this implies a perception of reality that is *indirect* rather than direct ("representative" realism as opposed to direct realism). Efforts to turn the mediating process itself into reality (perhaps the "adverbial" theory of perception) seem to me more promising, but only if reality is understood to mean something other than strict mind-independence. The idea that at least certain real properties are "relations" seems very helpful in this regard (see, e.g., Swoyer and Orilia, "Properties," §1.1.3, §7.4, as well as Sklar, *Philosophy of Physics*, 19–22, 73–82). See also Dretske, "Perception"; Crane, "Problem of Perception."

11. Contrast Descartes's distinction between external matter and interior mind with Aristotle's distinction between the *matter* out of which something is made (e.g., wood) and the *form* given to the matter by, for example, human beings (e.g., a boat).

12. Blackburn gives a summary of the "Galilean worldview" that supplanted Aristotle's organic view of the cosmos (Blackburn, *Oxford Dictionary of Philosophy*, 152, 287).

13. The Latin word *realis* (based on *res*, meaning "thing" or "object") was coined in the twelfth or thirteenth century by Roman Catholic scholars, the scholastics, who used it to refer to those identifying characteristics of things that came from nature and not from our way of understanding nature—not from, as we would say today, our minds.

These scholars were grappling with the possibility raised in the third century C.E. by Porphyry, and then in earnest by Roscelin de Compiègne (c. 1050–c. 1125) and his student, Peter Abelard (1079–1144), that language and the human mind might be responsible for all universals, including, as the debate unfolded, the "human nature" that distinguished humans from other creatures. For earlier scholastics and, well before them, for Plato and Aristotle, basic underlying properties (i.e., Plato's "forms"; Aristotle's "essences") were universals of nature and did not depend on the knower. Thus William of Champeaux, a contemporary of Roscelin and Abelard who sided with earlier scholastics and Aristotle, claimed in the traditional fashion that human nature was a naturally given, universal quality of all human beings. Roscelin, however, claimed that universals were merely *flatus vocis*, or words we breathe ("wind noises") and superimposed on nature by humans. Abelard fleshed out arguments for this idea, especially in terms of species distinctions. With such arguments being raised, the scholastics needed a term to designate universals that were in fact mind independent.

My history of the term *real* comes principally from Feibleman, "Real"; Dravid, *Problem of Universals*, 368–381; and De Wulf, "Roscelin." For our contemporary use of the word, in addition to standard dictionaries, see Blackburn, *Oxford Dictionary of Philosophy*, 319; or Forbes, "Reality," 775 (but note how Forbes's definition of reality as "independent of appearance" doesn't easily encompass human artifacts insofar as they might differ from just hunks of matter).

14. By "property," I mean what can be predicated of or attributed to things. Another way of saying this is that a property is what can be *instantiated* by things (Swoyer and Orilia, "Properties," §1.1; Bealer "Property," 752). To say that "red" is a property of all red things is to say that "red" is instantiated by all red things. Properties that can be instantiated by more than one thing are universals—what remains the same from one instance to the next. (Universals and particulars are the subject of the next chapter. The nature of properties is a central subject of this book.)

In this book, I treat relations as a kind of property (see Swoyer and Orilia, "Properties," Introduction, §7.4).

15. Goldman, *Science in the Twentieth Century*, 1:121.

16. For realism versus instrumentalism, see Rosenberg, *Philosophy of Science*, 91, 93. Bohr's version of instrumentalism seems to have been of a nuanced and perhaps weak variety (see, e.g., Faye, "Copenhagen Interpretation of Quantum Mechanics," §4, but also, e.g., de Muynck, *Foundations of Quantum Mechanics*, 187).

For another version of this mind-matter problem in the context of debates about the mathematical constants of nature in physics, see Barrow, *New Theories of Everything*, 112–116. Here, the problem is the origin of the constants that are fundamental in our theories of physics (e.g., Newton's gravitational constant, G; the speed of light in a vacuum, c). Do they show us reality itself or just our way of understanding an unknowable reality?

17. For properties in general, see prior note 14. For the relation between physical laws and observers, see chapter 3, note 3.

18. My distinction between everyday questions about reality and philosophical questions about the nature of everyday reality is, as far as I can tell, very similar to Heidegger's distinction between the ontical and the ontological. Heidegger, *Being and Time*, 11–15 [Macquarrie and Robinson 32–35].

19. In Brainard, *Reality and Mystical Experience*, a third-person account of reality is called "publicity based" and a first-person account is called "presence based." These original names are based on terminology to be presented later on.

Chapter 3

1. Gertrude Stein provides a good example of the type-token distinction in her well-known line "Rose is a rose is a rose is a rose." Were one asked to count the total number of words in this sentence (just the words themselves without regard to their meaning), there would be two possible answers. The result could be either three or ten depending on whether the request meant "words" as a type or a token (Wetzel, "Types and Tokens," §1.1).

2. Universal rules of change can, at least in physics, also be recast in terms of a second universal that does *not* change over time. When hydrogen combusts, for example, the form of the energy may change, but the total energy remains the same. These qualities that are involved in changing states but do not themselves change (e.g., total energy or total momentum) have given us some of our most important and wide-reaching laws of nature, such as the laws of thermodynamics (Barrow, *New Theories of Everything*, 22).

3. The most fundamental building blocks of physics are called *symmetries*. These symmetries not only underlie our laws of physics; they define what qualifies as a law. To qualify as a law of physics, an explanation of physical phenomena must have four symmetries. These symmetries define fundamental dimensions over which a law stays the same for all observers. A law is the same (1) over time. It is the same regardless of when an observation is made. $E = mc^2$ applies a billion years ago as well as today. (2) It is the same over space. It is the same throughout the universe regardless of where an observation is made. $E = mc^2$ applies in other galaxies besides our own. (3) It is the same regardless of the relative motion of observers. $E = mc^2$ applies regardless of how fast an object might be moving past us. (4) It is the same regardless of the angle from which an observation is made or how an object is rotated. $E = mc^2$ applies regardless of how we orient our experimental apparatus (Greene, *The Elegant Universe*, 167–170).

    Our current efforts to unify Einstein's general theory of relativity with quantum mechanics—that is, to unify our account of the very large with our account of the very small—implies still another symmetry. Laws of nature should be the same regardless of the scale of the observer. Thus a law of nature with this symmetry should remain the same regardless of whether we are observing a black hole or an electron. At present, such laws are not the same. We have two theories without, as yet, a satisfactory covering theory that would resolve the conflicts between them and work at any scale. (I will return to the subject of scale later.)

4. For a general discussion of universals in Indian philosophy, see Dravid, *Problem of Universals*; or Staal, *Universals*. For this subject in Chinese philosophy, see Graham, *Disputers of the Tao*. Fung, *History of Chinese Philosophy* also discusses the subject (e.g., see his index listings for "universal" and "chih"). For an anthology on the debate in the West since Plato, see Schoedinger, *Problem of Universals*.

5. For Mesopotamian and Near Eastern deities, see Jacobsen, "Mesopotamian Religions," 9:447–466 (for deity names, 450). Hindu deity names can be looked up in a Sanskrit dictionary to see their standard usage. (See also the chapter 8 discussion of Mesopotamian polytheism.)

6. Most of this summary of early Greek philosophy comes from Barnes, *Early Greek Philosophy*.

7. Barnes, *Early Greek Philosophy*, 18–19.

8. This philosophically pivotal time is Karl Jaspers's "Axial Age" (*Origin and Goal of History*).

9. Plutarch, *On the E at Delphi*, 392B, in Barnes, *Early Greek Philosophy*, 117.

10. For example, Plato's *Phaedo*, 74–75.

11. The fact that direct experience by itself doesn't show us universals was a main factor undermining logical positivism in the last century (see chapter 1 text box "Logical Positivism and the Limits of Science").

12. Schoedinger, *Problem of Universals*, 16–22. For a more general discussion of Aristotle's philosophy, see Cohen, "Aristotle's Metaphysics."

13. Isaac Newton insisted that the universal rules of nature were entirely independent of particulars. The laws governing any material object existed everywhere regardless of whether such things were present (Barrow, *New Theories of Everything*, 94–95).

14. The fact that perceptual recognition requires some degree of inference does not preclude it from being particular in respect to its here-and-now, interactive nature.

15. Not all universals are "lawgiving" in the sense implied here, and many people think that none are. While universals are by definition unchanging and capable of being various places at the same time (as well as often vague and/or indeterminate), only those universals that distinguish among naturally occurring classes of phenomena (what are called "natural kinds" or "sortals") can be said to be "lawgiving" (although some, including Buddhists, do not agree that natural kinds exist either). It is this specific kind of universal that I call a "fundamental nature" or an "archetype." While I think the concept of "atom" offers a plausible, easy-to-understand example of a natural kind, perhaps better examples from physics would be classes of subatomic particles, such as fermions and bosons, or the various species of quark, since the properties by which we identify atoms seem explainable in terms of subatomic particle properties.

16. The only exception to particulars always being made out of parts is the hypothesized "material simple" (see chapter 7, note 16).

17. The study of part-whole composition, "mereology," encompasses anything we might call a "part," and thus can include abstract entities (e.g., "rationality is part of personhood"). But the formal use of "part" in this broad sense is controversial, one important reason being that it doesn't seems reconcilable with their being universals (Varzi, "Mereology," §1). In any case, the distinction here between part-whole construction and class logic follows Körner, *Metaphysics*, 7–9. See also the discussion of set theory in chapter 11, "Mathematics and the Behavior of the Universe" (in mathematics, sets are usually consider to be a type of class).

18. My distinction between two kinds of causality reflects Aristotle's distinction between "efficient cause" and "formal cause" (Aristotle, *Metaphysics*, 1013a). In my context here, the cause-and-effect way one atom instance affects another corresponds roughly to Aristotle's "efficient cause," while the law-following behavior of atoms—how they obey universal rules—corresponds roughly to Aristotle's "formal cause." (Aristotle's "material cause" and "final cause" are mentioned in chapters 2 [note 11], 10 [note 4], and 12 [note 18].)

19. See Swoyer and Orilia, "Properties," §1.1. Richard Rorty's entry on "Relations, Internal and External" for the 1967 *Encyclopedia of Philosophy* is also helpful and still seems to apply (130–132). "Bare particulars" can be easily found in an online search.

20. Barnes, *Early Greek Philosophy*, 106–108.

21. In the context of realist versus antirealist debates, consider how Descartes's matter-mind distinction changed the historical distinction between a "thing in itself" and a "thing in relation to us" (chapter 2, note 5).

22. This collective-distributive ambiguity of the grammatical first-person plural also applies

to the grammatical second-person and third-person plurals. Like the example in the text, the following questions are also ambiguous: "Did all four of you write a book?" "Did all four of them write a book?"

Chapter 4

1. For Sanskrit words, I use conventional Western spellings when they exist; otherwise I use the Roman transliterations for the Sanskrit and Pāli (Pāli is the language of Buddhism's original scriptures). At the end of the book is a glossary that includes both transliterations and approximate pronunciations.

2. Nondual Hinduism has become a principal face of popular Hindu philosophy in the West, no doubt due in part to the similarities of its teachings to those of Christian, Islamic, and Jewish mystics. H. P. Blavatsky (1831–1891) and Henry Olcott (1832–1907), founders of the Theosophical Society, as well as Vivekananda (1863–1902), Paramahansa Yogananda (1893–1952), and others, were instrumental in bringing Nondual Hinduism to the West. More recently, its teachings have continued to be promoted in the writings of Ken Wilber, Ram Dass, Deepak Chopra, B. K. S. Iyengar, and many others.

3. Radhakrishnan and Moore, *Sourcebook in Indian Philosophy*, 16–27.

4. Note here the difference between Nondual Hinduism's account of reality and that of Spinoza, which has been likened to versions of Hinduism (as is easily found online). Similar to Nondual Hinduism's Brahman, Spinoza's God is the substance of the universe. For Spinoza, however, God is that which bridges the Cartesian gap and explains mind (i.e., thought) and matter (i.e., extension) as operating in parallel through God's orchestration (e.g., Garrett, "Spinoza," 871). Hindu philosophy, however, does not share Descartes's mind-matter divide. More like Plato, Aristotle, and other pre-Enlightenment philosophers, Hindus generally speak of the mind (Sanskrit: *manas*) as a faculty for receiving sense impressions (among other things), while matter (*prakṛti*) is the "stuff" or building material out of which all that we perceive is made. (For more on the meanings of mind and matter in Hinduism, search: [manas Sanskrit] and [prakṛti Sanskrit].)

5. *Tanakh: The Holy Scriptures: The New JPS Translation according to the Traditional Hebrew Text* (Philadelphia: Jewish Publication Society, 1985).

6. Bṛhadāraṇyaka Upanishad, I.4.3, in Campbell, *Masks of God*, 10, which includes a comparison of this account with that of Genesis. The creation story of the universe arising from a cosmic egg (*Hiraṇyagarbha*) appears in several places, such as the Rig Veda 10.121.1 (Radhakrishnan and Moore, *Sourcebook in Indian Philosophy*, 24), the Chāndogya Upanishad 3.19.1–4 (ibid., 65–66), and the Laws of Manu (Manu-smṛti) I.5ff. For a more extensive comparison between Indian and Judeo-Christian creation stories, see Campbell, *Masks of God*, 9–13. In spite of stories like this, the Dual school (*Dvaita-Vedānta*) of Hinduism takes a more Western approach, claiming that Brahman (who for them is Vishnu) is separate from us and *not* the substance out of which the universe is made (see, e.g., King, *Indian Philosophy*, 57).

7. While neither Hindu nor Buddhist philosophy shares the Cartesian divide between mind and matter (see this chapter's note 4), some Hindu schools do have distinctions among first principles that may seem similar to our mind-matter contrast. However, in these cases, reality is typically thought of as arising out of the *convergence* or *interplay* of mind and matter. The Sāṃkhya philosophical school, for example, views conscious selves

(*puruṣa*) and matter (*prakṛti*) as two principles of nature, but it conceives of reality as arising through the *interaction* of these two principles. Likewise with the Hindu Dual school
(*Dvaita-Vedānta*): it, too, conceives of God, human souls, and matter as separate players,
but again, reality arises through the relationships that aware beings (God and souls) have
with matter and each other (Raju, *Structural Depths of Indian Thought*, 308–321, 479).

8. The notion that the one supreme reality manifests in time and space as a threefold
   division among subject, object, and the relationship among them (Sanskrit: *tripuṭi*) is
   a central teaching of principally the nondual (Advaita-Vedānta) school. *Tripuṭi* means
   literally "triple base" or "triad" and is also used in other contexts besides knowledge (e.g.,
   lover, beloved, and the love between them; search: [Hinduism tripuṭi]). This same idea
   can also be found in at least Jewish philosophy (Maimonides, *Guide for the Perplexed*,
   1:68 [Friedländer 100–101]).

9. It may seem as if Hindus have made a "category mistake"; because all aware selves share
   the same fundamental nature (i.e., aware selfhood), it follows that this fundamental
   nature must itself be an aware self. But a category mistake like this presumes a certain
   conception of reality, one involving categories that, at least in the West, originate with
   Aristotle. Hinduism in fact has a philosophical school, the *Vaiśeṣika*, that uses categories somewhat similar to those of Aristotle. However, in the schools of Hinduism
   emphasized here (esp. Advaita-Vedānta, the nondual school), reality doesn't match
   these categories, and the universal, at least conceptually, is *more fundamental* than the
   particular. In other words, particularity begins with and arises from the universal.

   This idea that Brahman transcends the universal-particular distinction is echoed
   in monotheist philosophy (though in a way appropriate to its dualist approach). In his
   account of how God is related to living creatures, Thomas Aquinas writes, "God is not
   related to creatures as though belonging to a different genus, but as transcending every
   genus, and as the principle of all genera" (Aquinas, *Summa Theologica*, I, Q. 4, Art. 3
   [Pegis vol. 1, 41]).

10. *Panentheism* is a Western term for God conceived of as both distinct from the universe
    and yet also its underlying substance. It also has been frequently used to refer to Non-
    dual Hinduism's concept of Brahman. However, as I'll discuss in chapter 8, Brahman
    should not be thought of in terms of our usual Western idea of substance. Hartshorne
    and Reese's *Philosophers Speak of God* remains one of the main texts on this subject.

11. Bhagavad-gītā, 13.13–17, 15:7–9.

12. Māṇḍūkya Upanishad 27.

13. Brahman may be thought to have the qualities of sat, chit, and ānanda; nevertheless,
    Nondual Hinduism conceives of Brahman as beyond conventional understanding. Any
    effort to make Brahman more concrete and understandable turns Brahman into *Ishvara*
    (*Īśvara*, literally, "lord of the universe"), a personal God very much like that found in
    Western monotheism.

14. Deutsch, *Advaita-Vedānta*, 9–14. Deutsch writes that "Bliss (ānanda) points to the prin-
    ciple of value . . ." (10).

15. Hinduism's unapologetic association of Brahman with the archetypal human is espe-
    cially evident in its concept of *puruṣa*. While the term literally means "person" or "man,"
    it often also refers to the original or eternal person, and in Nondual Hinduism, puruṣa is
    identical with the true self, Ātman, and thus to Brahman (e.g., Bhagavad-gītā, 15.16–19).
    Hinduism's use of puruṣa is somewhat similar to *Adam Kadmon* in the Kabbalah and
    more generally to *God Anthropos* (as scholars call it) in Gnostic traditions. Greek and
    Roman gods are other familiar examples of this association of deities with the archetypal

qualities of humankind. (Filoramo, *A History of Gnosticism*, 58, 87. See also McGinn, "Love, Knowledge, and *Unio Mystica*," esp. 75–81. For a more extensive discussion of Adam as the primordial human, see Idel, *Kabbalah*, 112–199.)

16. As in other religions, a Hindu's life experiences are presumed to be affected by spiritual practices that he or she is responsible for performing. In Nondual Hinduism, these practices are concerned with discovering one's identity with Brahman and behaving in a way that reflects this identity and leads to enlightenment (the subject of the next section).

17. Chāndogya Upanishad 6.12.13, in Radhikrishnan and Moore, *Sourcebook in Indian Philosophy*, 69.

18. Nondual Hinduism's conception of enlightenment is also tied to the logical principle of noncontradiction. Nondual Hindus conceive of reality as having four levels gradated according to their logical coherence. The lowest level, "insignificant being" (*tucchasattā*), comprises those things that are contradictory in themselves and thus never thought of as real—a "married bachelor," for example, or a "square circle." The next level, "apparent or illusory being" (*prātibhāsikasattā*), comprises those things that appear real but can be contradicted by other observations, such as when we mistake a rope for a snake or a post for a person. The next level, "pragmatic or empirical being" (*vyāvahārikasattā*), refers to what cannot be contradicted by any other observation and is in fact true for all—what we might today call empirical reality. If the object someone sees is truly a person, then it is a person regardless of whether someone else mistakes it for a post.

   This "empirical" reality is not, however, ultimate reality. Suggestive perhaps of the limitations of empirical science, this level is itself contradicted by and understood in terms of Brahman, the ultimate reality, which is itself the root of logic and of understanding (*pāramārthikasattā*) (Raju, *Structural Depths of Indian Thought*, 389–390; Deutsch, *Advaita-Vedānta*, 15–26; Brainard, *Reality and Mystical Experience*, 165–167).

19. Śaṅkara, *Vedānta Sūtras*, II.i.33 [Thibaut 533].

20. *Rāja Yoga*, or "eight-limbed" (*ashtanga*) yoga, comes from the teachings of Patañjali (dates vary from 400 B.C.E. to 400 C.E.). Vivekananda, for example, presents Rāja Yoga as a separate path (Vivekananda, *Rāja-Yoga*). See Smith, *World's Religions*, 41–50, for an accessible introduction to Rāja Yoga.

21. Mother Teresa, *A Simple Path*, 7–15.

22. Taittirīya Upanishad, 2.1.3–2.5.2 (Nikhilananda, *The Upanishads*, 266–268; see also Nikhilananda's endnote).

23. For the "inexplicability" of Brahman and maya, see Brainard, *Reality and Mystical Experience*, 169; or Raju, *Structural Depths of Indian Thought*, 386–388, 392.

24. Ibid.

Chapter 5

1. Aristotle, *Metaphysics*, 1010a11.

2. Guthrie, *A History of Greek Philosophy*, Vol. 3, 194. Also search: [Gorgias "On the Non-existence"]. For Skepticism, see chapter 1, note 9; and chapter 2, note 6.

3. For more on the phenomenological self, the self that is a generic self (an "I" that is a "we"), see Husserl, *Ideas*, §27–30 ("The Thesis of the Natural Standpoint") and also the explanation of this standpoint given by Kohák, *Idea and Experience*, 29–35.

4. Much of this section's history of universals comes from Dravid, *Problem of Universals*.

5. Descartes, *Meditations on First Philosophy*, 313.

6. Husserl gives his most thorough discussion of phenomenology in *Ideas*. For Husserl, the decisive factor of his method lies "in the absolutely faithful description of that which really lies before one in phenomenological purity, and in keeping at a distance all interpretations that transcend the given" (Husserl, *Ideas*, §90). For his analysis of how we imaginatively adopt other points of view, see *Ideas*, §4, §41, and the index entry for "Fancy (Phantasie)." For Husserl's indebtedness to Descartes, see Husserl's introduction to *Cartesian Mediations*. For the difficulties with solipsism that infect Husserl's phenomenology, see, for example, Ricoeur, *Husserl*, 123ff., 142, 174, and Edie, *Edmund Husserl's Phenomenology*, 76–77. Husserl was aware of and discussed these difficulties, for example, in his Fifth Meditation in *Cartesian Mediations*. For Heidegger on how our first-person view is a "being-in-the-world," see Heidegger, *Being and Time*, 52–53 [Macquarrie and Robinson 78–79].

7. That many writers link Buddhism with phenomenology is clear from a search for the two terms. Also, Robinson and Johnson's textbook on Buddhism speaks of the Buddha's Four Noble Truths as "best understood not as the content of a belief, but as phenomenological categories, a framework for viewing and classifying the processes of experience as they are directly present to the awareness" (*Buddhist Religion*, 35). For a specific overview of what phenomenology offers to the study of Japanese Buddhism, see Shaner, *Bodymind Experience*, 2–5. For a phenomenological analysis of Yogācāra Buddhism, see Lusthaus, *Buddhist Phenomenology*. See also chapter 6, note 5.

8. Husserl would likely have approved of my including an imaginary cat in my phenomenological account; see prior note 6 and its references for "Fancy (Phantasie)."

9. For Husserl on the "natural standpoint" and "bracketing," see Husserl, *Ideas*, chap. 3.

10. Consider space and time from the first-person view as compared to the third-person view. From the third-person view, space and time serve as impartial, mind-independent mediums onto which we map all the here-and-now events that ever were, are, and will be. Columbus crossed the Atlantic in 1492; I'm sitting here writing this in 2016 in Pennsylvania (which is on planet Earth, which is in the Milky Way galaxy, etc.). From the first-person view, however, time and space are not at all impartial or mind independent. At the center of all time and space is my own subjective here and now with other here and nows ranged about and separate from me. Time and space are for me what distinguishes my present here and now from all other here and nows. I am always right here and now. Columbus's own here-and-now boat landing is for me a long time ago. Whatever might be going on in Tokyo right now seems very distant from me in space. (For Husserl on phenomenological time, see Husserl, *Ideas*, §81.)

11. The concept of "public" will be further refined in later chapters; it will not oppose "private" in this fashion.

   It is unclear to me whether consciousness for Husserl distinguishes among different ways of being aware such that essences ("noemas") might vary according to the kinds of awareness they are associated with (Husserl, *Ideas*, §152–153). However, the earlier philosophical idealist George Berkeley does seem to do this (chapter 6, note 10).

12. The object I see before me as a "desk" may not be a "desk" to a cat or spider, but there is still that about it which allows me to point to it and consider it from other points of view besides my own. There is some "it" or "thing" that remains sufficiently the same over our different views in order for "it" to appear to each of us in whatever different ways it does (see, e.g., Husserl, *Ideas*, §40, §149–150).

No doubt phenomenological analyses lead eventually to this perceptual substrate of our first-person awareness of our self and the world, but it has proven very difficult to define it meaningfully. Husserl originally called it *hyle* after the Greek word for matter in general or "stuff" but later abandoned the term (Lusthaus, *Buddhist Phenomenology*, 14). Earlier, Berkeley denied the existence of matter as separate from perception but substituted "an omnipresent eternal mind, which knows and comprehends all things, and exhibits them to our view" (*Three Dialogues*, 3rd Dialog, 674). These days, philosophers are more apt to speak of the "conditions for first-person awareness," "sensory substrate," "background conditions," or "conditions of the possibility of intentionality." It seems also to be at this point where phenomenology overlaps contemporary research in not only the physical sciences but also the cognitive sciences. See Lusthaus, *Buddhist Phenomenology*, 13–14; Smith, "Phenomenology," §3.

13. For what we perceive as arising from our participation in a world over time (both collectively through evolution and individually in our own lives), see Grim, *Philosophy of Mind*, 188–205. Grim, for example, speaks of J. J. Gibson, who has argued that what is presented to us in our perceptual experience are always "affordances," possibilities for action in the world. Grim finds considerable merit in Gibson's approach, although he disagrees that all properties can be explained in this way (189–192).

14. Grim, *Philosophy of Mind*, 313.

15. For a discussion of consciousness in animals, see Blackmore, *Consciousness*, 166–179.

16. Higher-order (HO) theories of consciousness are varieties of functionalism, and the predominant current theories of consciousness are functional theories. Functional theories of the mind say that mental states are distinguished not by specific physical states of the brain but rather by specific functions, which need not reflect specific brain states. A light switch, for example, is defined not by what it's made of or by its size, shape, or how it works but rather by its use for turning a light on and off. Similarly, a functional theory of consciousness has less to do with our brain's construction than with what is accomplished by conscious awareness. Another frequent analogy likens consciousness to computer software that runs on the hardware of the brain.

   For more on functional theories of mind and higher-order theories of consciousness see Grim, *Philosophy of Mind*, 88–98, 381–392; Blackmore, *Consciousness*, 47–48; Siewert, "Consciousness and Intentionality," §6; Droege, "Higher-Order Theories"; Gennaro, "Consciousness," 4:b; Block et al., *Nature of Consciousness*, 719–806; Carruthers, "Higher-Order Theories of Consciousness."

17. For more on this distinction between conscious and unconscious awareness, see Baars, "Contrastive Phenomenology." Many studies of brain damage have also shown our capacity to perceive unconsciously what is hidden to our conscious awareness (e.g., "blindsight," "spatial neglect," and "prosopagnosia"). See Shallice, "Modularity and Consciousness," 256–261; Bisiach, "Understanding Consciousness," 246–247; Farah, "Visual Perception and Visual Awareness," 210–213. See also "Damaged Brains," in Blackmore, *Consciousness*, 257–270.

Chapter 6

1. Thich Nhat Hanh, *Touching Peace*, 1–2. Reprinted with permission from Parallax Press, Berkeley, CA, http://www.parallax.org.

2. The original teachings of the Buddha and his disciples are recorded in the Pāli Canon (named after the Pāli language of the Buddha's everyday discourse). Also called the "three baskets" (Pāli: *Tipitaka*; Sanskrit: *Tripiṭaka*), it is divided into three collections. The first collection gives the rules of Buddhist monastic life, the second holds the discourses and other teachings of the Buddha, and the third holds the commentaries of the monks on Buddhist psychology and philosophy.

   In discussing Buddhist philosophy, I use the conventional Western terms if they exist; otherwise, I will use Pāli transliterations for most of the terms and Sanskrit transliterations for some of the later concepts. The glossary gives both Pāli and Sanskrit for all terms.

3. For more on these cultural roots of Buddhism, see Harvey, *Introduction to Buddhism*, 9–14; or Ludwig, *Sacred Paths of the East*, 82.

4. Perhaps the closest Buddhism comes to a universal formula for characterizing its core components is the phrase "Three Jewels" or "Three Treasures." The first jewel is the *Buddha*, which can mean, among other things, either the historical Buddha or the Buddha as an idealized exemplar of enlightenment. The second is the *Dhamma*, or Truth, that he taught and that is realized in meditation. The third jewel is the *Sangha*, or community that follows the Buddha's teachings.

5. The Buddha was famously silent on most philosophical issues (*Cūla-Māluṅkya-Sutta*, no. 63 of the *Majjhima-Nikāya*, in Rahula, *What the Buddha Taught*, 13–14). Consider, in this regard, how similar his silence on metaphysical matters is to phenomenological "bracketing," or the putting aside of philosophical questions other than those that pertain to first-person experience.

6. Smith, *World's Religions*, 115–117; Harvey, *Introduction to Buddhism*, 47–53; Robinson and Johnson, *Buddhist Religion*, 37–39.

7. Perhaps it is difficult to understand interdependent arising as a reaction *against* universals when it seems to be itself a universal, an enduring pattern of cause and effect. In Siddhartha's time (and in Hinduism today), existence referred to what endured from moment to moment (much like Aristotle's *ousia*, substance). In the context of our everyday world, the things that made up our life, like people, trees, houses, and rivers, existed while they were part of the world and ceased to exist when they died, disintegrated, dried up. The Buddha's concept of interdependent arising was in part a reaction against this idea that something either existed or didn't exist at all. In its place, he introduced the idea of the world as a here-and-now *process* of sentient interaction that produces the seemingly enduring things that we perceive.

   Pratītya-samutpāda is variously translated as "interdependent arising," "dependent coarising," "conditioned genesis," "conditioned arising," "co-dependent origination," "dependent origination," or just "interdependence" (Laumakis, *Introduction to Buddhist Philosophy*, 105; Edelglass and Garfield, *Buddhist Philosophy*, 9). Key Buddhist sources are *Mahānidāna Sutta* (part 1), *Saṃyutta Nikāya* (vol. 2), and *Visuddhimagga* (ch. 17). Information about interdependent arising is readily available online. Also Robinson and Johnson, *Buddhist Religion*, 23–29; Hart, *Art of Living*, 47–51; Kalapahana, "Pratītya-Samutpāda," 11:484–488.

8. For the Buddha, both the world that we're conscious of and the path to our salvation are inseparable from our sentient (i.e., aware) bodies: "Within this fathom-long sentient body itself, I postulate the world, the arising of the world, the cessation of the world, and the path leading to the cessation of the world" (*Aṅguttara-Nikāya* 4:45, quoted by Rahula, *What the Buddha Taught*, 42).

9. According to the Pāli Canon, each moment of experience involves aggregates or collections (*khandas*, Sanskrit: *skandhas*) of *dhammas* (Sanskrit: *dharmas*). The word "dhamma" in this sense refers to the elementary particles of experience and is not to be confused with other meanings dhamma and its Sanskrit equivalent, dharma, have for both Buddhists and Hindus, such as "truth" and "right action." Dhammas are also not the same as the elementary particles of Western physics. Dhammas are constituents of first-person experience, not third-person matter: "Basically, the *dharma* [i.e., *dhamma*] theory provides an explanation of how the universe functions within the context of a sentient life, in particular a human flux, for it is human life that Buddhism is concerned with" (Skorupski, "Dharma," 4: 334). Some dhammas make up material objects, but others make up the perceptual process and even consciousness—all considered from the standpoint of the one doing the experiencing. For nonencyclopedic discussions of the khandas (skandhas), see Rahula, *What the Buddha Taught*, 20–26; Harvey, *Introduction to Buddhism*, 49–53, 84; Herman, *Introduction to Buddhist Thought*, 154–156, 176–177; King, *Indian Philosophy*, 116–120.

10. George Berkeley probably never posed the well-known question about a tree falling in the woods, and in any case, the alarm clock example is more apt for my purposes here. But Berkeley does discuss how properties change according to one's point of view and even according to the type of sentient being (*Principles of Human Knowledge*, 2 §6 [Chan 591]; *Three Dialogues*, 1st Dialog at middle [Chan 645–646]). He notes, for example, that while the "foot" of a very small insect would be extremely tiny to a human, it would be much larger to the insect itself and even larger to creatures of still smaller size (ibid., 1st Dialog [646]). Nevertheless, Berkeley and the Buddha view perceptual awareness differently. For Berkeley, "*esse est percipi*," or "to be is to be perceived (or to perceive)" (Berkeley, *Principles of Human Knowledge*, 2 §3 [Chan 590]). For the Buddha, even mind and perception arise through here-and-now causal interaction.

11. What about physical devices developed by humans to work when no human is around? Such devices rely on strictly physical properties unless other creatures besides humans are directly involved. Thus when a circuit breaker trips on the electric panel of my house, the only properties needed to explain what happens are physical ones. Properties that pertain only to humans (e.g., that it is a device to shut down an overloaded electric circuit, that it was built and installed by humans, that it helps prevent house fires, that it is built out of thus and so physical components) are entirely latent unless a person is around. The same holds for a trap set to catch a stray cat, except that in this case the properties needed to explain how the trap catches a cat must include the properties of cats as well (e.g., how a cat is lured into the trap and trips the door shut). Again, these species-specific properties come into play only when a cat is around to respond to the device as a "cat."

12. Warder, *Indian Buddhism*, 98.

13. About a century after the Buddha's death, Buddhism started splintering into a number of sects. One of the main sources of doctrinal disagreement was how to reconcile the Buddha's teaching of impermanence and no-self with the concepts of causality and karma: causality implies an enduring, predictable orderliness in the behavior of things, and karma implies an enduring person who is subject to karma and may attain salvation by elimination of karma. Karma without an enduring person seems, on the face of it, impossible: without an enduring self, how do the merits and demerits of today's behavior carry forward into future repercussions? If the "I" of today is not the "I" of tomorrow or the next life, how can there be karma at all? And why should that "I" of the moment be virtuous?

The first and most important of these schisms produced today's largest collection of sects, Mahayana Buddhism, whose responses to these questions I'll discuss in a later section. About a century or so later (c. 280 B.C.E.), the breakaway group this time was the Pudgalavādins (Vātsīputrīya school), or the "Personalists." To reconcile anatta (no-self) with karma, they posited a *pudgala*, an insubstantial version of "enduring person" (although it seemed to Hindu and other Buddhist schools to be just a version of the ātman in disguise). A few decades later, another group arose, the Sarvāstivāda, or the "everything exists" school. Their solution was to posit not the reality of the person but rather *sarvam asti*, or a theory of time in which everything exists in some sense in the here and now, regardless of whether present, past, or future (which has an interesting similarity to the "holographic principle" proposed by physicists as part of a theory of quantum gravity: Smolin, *Three Roads to Quantum Gravity*, 169–170).

For more on the problems of Buddhism's first-person philosophy and the fractious disagreements in early Buddhism over how to handle these problems, see Brainard, *Reality and Mystical Experience*, 91–97. For an in-depth analysis of the related problem of the Buddhist unconscious, see Waldron, *Buddhist Unconscious* (e.g., how do Buddhists explain the persistence of self through periods of unconscious like sleep; why don't we die instead? [Waldron, *Buddhist Unconscious*, 79]).

14. For nirvana being beyond logic and reason, see Rahula, *What the Buddha Taught*, 42–44.
15. Arthur Schopenhauer associates his philosophy with Buddhism (Blackburn, *Oxford Dictionary of Philosophy*, 343), but compare Schopenhauer's pessimism to the Buddha's optimism and the positive joy of nirvana.
16. For a discussion of nirvana as the Buddhist goal in life, see Rahula, *What the Buddha Taught*, 35–44.
17. Much of what I've discussed so far concerns the Buddha's teachings about wisdom (*paññā*). The Buddha also talks about compassion, or *karuṇā* (Rahula, *What the Buddha Taught*, 46), which came to play an even more important role in later Mahayana Buddhism (discussed next), with its concept of *bodhisattva* as the ideal of enlightenment. One way to think about interdependent arising is that it characterizes us as reality creators in our interactive relationships with the world. It puts us in the reality-creating business *cooperatively* with others. Thus to practice compassion—to embrace other points of view as one's own—is, in part, to practice this truth by seeing everyone as participants in the interdependent arising of our shared world.
18. While Buddhist and Hindu accounts of reality often differ sharply, there are many underlying similarities in their characterizations of life's goal. For one example, in Buddhism, one finds nirvana by moving beyond the desires that bind us to samsara. In Nondual Hinduism, the fifth and final cloak around Brahman that conceals our true nature is that of ānanda, or bliss, the principle of value whereby we make our choices in life.
19. The distinction between direct experience and thought and ideas is, for example, central to David Hume's philosophy. In his *Enquiry concerning Human Understanding*, he notes that our thoughts may mimic direct experience but that direct experience has a "force and vivacity" not normally found in thoughts and ideas (Hume, *Enquiry*, beginning of section 2 of "Of the Origin of Ideas" [Chan 705–706]).
20. The philosopher Jacques Derrida (1930–2004) writes explicitly about presence and its ineffability. He claims that in our efforts to understand the nature of reality, we have access only to what is present to us; however, the nature of this "being-present-to-us"— the "being of presence"—never reveals itself as such. It can never itself be translated into

a universal. It is this opacity of presence to its own nature that Derrida exploits in his method of philosophical "deconstruction," which purports to expose all philosophical foundations as rooted in language and culture. See Sarup, *Post-Structuralism and Post-modernism*, 37–38; Derrida, *Margins of Philosophy*, 5–9.

21. Blackmore, *Consciousness*, 25–28 (esp. Frank Jackson's often-cited "Mary's room" argument for the existence of qualia). For arguments against the existence of qualia, see Dennett, "Quining Qualia."

22. By "in some way actually aware of," I mean "consciously or unconsciously aware of." The issue here is here-and-now awareness of any kind, not just conscious awareness.

23. See prior note 20 for Derrida on this subject.

24. Throughout Buddhism, śūnyatā (or śūnya) carries the sense of "emptiness" or "void"; however, its meaning varies some from school to school. For Nāgārjuna, for example, śūnyatā contrasts with what can be signified and indicates that "ultimate reality" is less an objective referent than a reality empty of meaningful signification (Nāgārjuna, *Mūlamadhyamakakārikā*, 22:11 [Kalupahana 307–308]). In Zen, Dōgen uses the word to indicate an empty reality that is paradoxically both illusory and lucid (Dōgen, *Flowers of Emptiness*, chap. 14 [Kūge] and 182n7). The Gelukbas of Tibetan Buddhism describe this emptiness as a "nonaffirming negative" (*prasajya-pratiṣedha, med dgad*), a negation of a thing that does not support the existence of something else. For example, to say an object is not red merely asserts that it is some other color besides red; to say that it is śūnya asserts that it is not red nor any other color, nor even not non-red (Klein, *Knowledge and Liberation*, 27–28, 151–152). See also Herman, *Introduction to Buddhist Thought*, 278, 304ff.; Dravid, *Problem of Universals*, 264–265. The story of the Śūnyavādins in the text comes from the *Encyclopedia of Eastern Philosophy and Religion*, 330.

25. In perhaps his most famous work, *Mūlamadhyamakakārikā*, Nāgārjuna criticizes the logic behind a broad array of first principles found in the philosophy of his time. He does not, however, similarly criticize the process of interdependent arising (which he reviews in chap. 26). See also Brainard, *Reality and Mystical Experience*, 103–107.

26. The Yogācāra (Sanskrit for "yoga practice") school elevates the mind/consciousness to a first principle, a philosophical strategy that allows them to account for the enduring selves on which karma, virtue, and enlightenment (as well as what I've call "latent" properties) depend. The properties that define sentient beings are thought of as karmic "seeds" (*vāsanā*) that are preserved in a universal "storehouse consciousness" (*ālaya-vijñāna*), where they ripen and eventually blossom into a new incarnation.

While traces of Yogācāra thought can be found even in the Hindu Upanishads, the Buddhist school begins principally with Vasubandhu of the fifth century C.E., who taught a doctrine very close to today's phenomenology. Yogācāra also grapples with some of the same issues that concern phenomenology, such as the problem of how we know there are other minds if all we know is the content of our own minds (Herman, *Introduction to Buddhist Thought*, 333).

Chapter 7

1. See chapter 2, note 10.

2. Kelly, "Evidence," §4; Wood, "Publicity." In the section where Kelly discusses publicity, he also discusses Carnap's evolving views as a logical positivist and, in this context, the

problem that logical positivists had explaining the "public" quality of empirical evidence insofar as it was based on sense data (see chapter 1 text box "Logical Positivism and the Limits of Science").

3. For an interesting alternative to the concept of presence, one proposed by the thirteenth-century Roman Catholic scholar Duns Scotus, search for [haecceity]. The concept of haecceity was later transformed by Leibniz into a cornerstone of his philosophy of monads.

4. "Thereness" in this context is meant to connote not a spatial location, a "there" versus a "here," but rather a presence versus an absence, a "being there" versus a "not being there" (though this dichotomy is misleading insofar as absence might itself be a kind of presence).

5. For more on the idea that reality itself is made, at least in part, of possibilities (what I'm calling "capabilities for presence"), see, for example, Peirce, *Philosophical Writings*, 300–301; or Whitehead, *Process and Reality*, 65ff.

6. "Capability for presence" refers here roughly to what we often mean by "existence." For example, to say that the Eiffel Tower exists now but didn't 150 years ago is roughly to say that this particular configuration of matter is a possible presence in human awareness today but wasn't 150 years ago. To say that space travel exists while time travel doesn't is roughly to say that space travel is something that some humans are capable of experiencing while time travel is not. The principal difference between the two terms is that "existence" can be (and, it seems to me, usually is) grounded in third-person philosophy, while presence is grounded here in first-person philosophy. Existence in this third-person sense is a quality that a thing itself has irrespective of its relationships to other things, which isn't the case with presence.

I use "capability for presence" instead of "existence" in my definition of reality for three reasons: (1) existence has other meanings that might be confusing here, (2) I find that the concept of presence and its connotations pair up better with the concept of publicity, and (3) I need to preserve the word "existence" and its conventional meanings in order to have them available for discussing reality and presence.

7. See, for example, Rosenberg, *Philosophy of Science*, 32–35. For one popular way of dealing with counterfactual conditionals as well as with the alternate realities of quantum mechanical probability waves, see the "many-worlds interpretation" of quantum mechanics (chapter 11, note 31).

8. The "presence" of another person, snake, or rope in first-person awareness no doubt seems different in kind from the presence of an electron or photon in a material interaction. One involves awareness; the other does not (at least in our usual sense of awareness). Nevertheless, as I'll explain in later chapters, I think these two kinds of presence are more similar than one might initially think.

9. For a discussion of everyday reality in Indian philosophy (i.e., samsara) as characterized by both publicity and presence, see Brainard, *Reality and Mystical Experience*, 70–72, 129.

10. See chapter 2, note 10.

11. The debate over the extent to which *phusis* implies "self-sufficiency" is evident in our interminable debates over free will.

12. Barnes, *Early Greek Philosophy*. 19–20. See also chapter 2, note 5.

13. Things with *phusis*, things with their own nature, have been central in debates over the reality of universals. Recall, for example, how our word "real" originated in debates over the mind-independence of universals, especially the species traits that distinguished one creature from another.

14. For the philosophical difficulties of causation, see, for example, Kim, "Causation," 127; or Blackburn, *Oxford Dictionary of Philosophy*, 59. Causation also posed difficulties for

Hindu philosophy (King, *Indian Philosophy*, 198–228), difficulties that the Buddhist philosopher Nāgārjuna exploited in the first chapter of his famous attack on all metaphysics, his *Mūlamadhyamakakārikā*.

15. A cause not caused by anything else—a "first" cause—seems difficult to find. For example, we might trace a broken window to a person throwing a rock, but why should we consider the person to be the agent in this case when a person's choices and actions seem to have their own prior causes?

   This question confronts us not only with the issue of free will but also with an infinite regress of causes without any apparent agent at all. For Thomas Aquinas, this infinite regress offered proof that there must exist an ultimate first cause or agency at the beginning of this causal chain—a proof for Aquinas of the existence of God (Aquinas, *Summa Theologica*, I, Q. 3, Art. 3 [Pegis vol. 1, 22]). Islamic philosophy takes another approach. It proposes that it is *between* causal moments that we find God's agency; God is that which binds causal moments together (see chapter 8, note 29). (See also Keil, "Making Something Happen.")

   The question of whether a cause can be in some sense intentional or self-willed pertains to the free will versus determinism debate, a subject discussed in subsequent chapters (see, e.g., the text box in the next chapter, "Free Will, Evil, and Accounts of Reality").

16. In the case of matter, we would expect to find individual agency in particles of matter having properties that didn't come from anything smaller (however, see chapter 10, note 13). Possible candidates include electrons, quarks, photons, and the strings of string theory. Such particles are called "material simples" if they are presumed to have no substructure (as per Kane, "Physics beyond the Standard Model," 71). (However, "material individuals" might be a better term than material simples, since an absence of substructure is arguably not entailed by individual agency ["Reality as Emergent" in chapter 11; Gilmore, "Location and Mereology," §5.3–5.5].)

17. This distinction between individual and collective agency affects all aspects of our lives and is a vast subject of contemporary research. It's central to politics and law, where collective interests are decided, codified, and enforced as rules regulating the behavior of constituent individuals. It underlies economics, where the desires of individuals for goods and services generate collective forces that affect everyone. It pertains to ethics as collective guidelines for individual behavior. It also pertains to physics, where, for example, a beam of light can be thought of as the collective behavior of many individual electromagnetic particles.

18. Recall Aristotle's distinction between material and formal cause (chapter 2, note 11; also chapter 10, note 4).

19. The idea that the universe itself has built-in groups distinguished by properties (i.e., "natural kinds" or "sortals") of which they themselves are the source is controversial. Also consider the debates over the theory of "emergence": are "life" and "consciousness" in fact properties of the universe that have emerged from but cannot be explained by or reduced to material properties, or are they perhaps merely levels of complexity we don't understand? (See chapter 11, "Reality as Emergent.") In my context here, note that these are debates over *where* such properties originate—whether or to what extent they come from the human mind.

20. If the color of emeralds is a publicity rather than merely a universal, then their color is a predicate to be understood in the context of physical uniformities maintained by matter over space and time. Note that this lawlike behavior of matter is not another piece of data that I have surreptitiously sneaked in. It's an aspect of the original

data, an aspect that we in our everyday lives always implicitly include. This proposal leaves open for now the question of how awareness and matter accomplish these uniformities.

Again, "grue" doesn't intend to refer to anything actually occurring in nature—for example, emeralds that actually change color from green to blue at a certain time. It intends only to show how it's possible to come up with a host of predicates (Goodman gives others as well) that accurately describe our experience but that we would usually reject for intuitive reasons that formal logic doesn't currently seem to encompass. Goodman uses these predicates to ask where inductive logic gets its warrant for "lawlike" statements about our world, statements that forecast how something (e.g., a piece of copper, an emerald) will behave in circumstances beyond those we've already observed (Goodman, *Fact, Fiction, and Forecast*, 73). Quine's answer to Goodman's problem, for example, is "natural kinds" (Bird and Tobin, "Natural Kinds," §1.2.1), which is notably somewhat similar to what I've described as "collective agency."

Goodman gives the following definition for "grue": "The predicate 'grue' . . . applies to all things examined before *t* just in case they are green but to other things just in case they are blue" (Goodman, *Fact, Fiction, and Forecast*, 74). The grue problem is easily found online. Apart from Goodman's *Fact, Fiction, and Forecast*, my main sources were Vickers, "Problem of Induction"; Johnson, "Grue Paradox"; and Stalker, *Grue!*

21. See prior note 16.
22. Universals are publicities with class instances differentiated in space as well as time. Particulars are publicities with class instances (or, perhaps better, "moments") differentiated, if more than one, only in time.
23. In considering the implications of observers not just observing but also interacting, recall the importance of "observers" in defining a law of physics (chapter 2 section "The Concept of 'Reality'").

Chapter 8

1. James, *Varieties of Religious Experience*, 407.
2. For a general discussion of gods and goddesses, see Ludwig, "Gods and Goddesses," 6:59–66, and Wright, *Evolution of God*, 1–98. Also helpful in my context here is Geertz, *Interpretation of Cultures*, 87–125, as well as Gold, "Consecration," which discusses how everyday objects become sacred or holy. For an old essay (I assume there are more up-to-date examples) on how the use of gods in pre-science theoretical thought parallels today's use of concepts in science, see Horton, "African Traditional Thought." For a critique of Horton and a discussion of how these two sorts of rationality might differ, see Ulin, *Understanding Cultures*, 63–88. (But note in his critique the importance Ulin gives to cultural and linguistic factors in explanations using deities [76].)
3. Not surprisingly, the line between human individual agency and divine collective agency can blur for persons who have exceptional collective importance. Tribal chiefs, kings, important ancestors, shamans, saints, founders of religions, and the like are often viewed either as themselves deities or as bridges to the spirit realm (Wright, *Evolution of God*, 46–69).
4. Principally, Jacobsen, "Mesopotamian Religions," 9:447–466; Black and Green, *Gods, Demons, and Symbols*; Jordan, *Encyclopedia of Gods*; Fulco, "Baal," 2:31–32. Robert

Wright points out that, in a certain sense, the "hunter gather religion . . . doesn't exist" (Wright, *Evolution of God*, 19) because their deities are not something apart from the everyday dynamics of their world.

5. I've highlighted here the importance of deities as agents of reality in the sense that they point, as best a community can determine, to where universals come from—what is responsible for the collective order (and disorder) of the world. The anthropologist Clifford Geertz points to two related ways deities served a community's collective understanding of reality. On the one hand, deities are "models of" reality; they help people understand reality by embodying its basic properties. They put a face on the properties that characterize all sheep or all grain or all fire. On the other hand, they are "models for" reality; they shape collective thought and behavior so as to construct a certain shared worldview. They are concepts shared by a community that allow the group to view the world out of collective eyes and construct for themselves a relatively consistent view of nature (Geertz, *Interpretation of Cultures*, 95).

6. Many writers have commented on how the deities of old are similar to what we now think of as aspects of our psyches. Perhaps the best known of these writers was Carl Jung, who advocated a theory of archetypes that explicitly recall the deities of our forebears (Jung, *Two Essays on Analytical Psychology*, 69–70, 94). Jung also famously proposed that the God of monotheism was perhaps more accurately thought of as the "collective unconscious" (ibid., 71).

7. For the development of Jewish monotheism, see especially Wright, *Evolution of God*, 99–244. See also Halpern, "Monotheism," 524–527; Sperling, "God in the Hebrew Scriptures," 1–6; Armstrong, *History of God*, 1–39. My dates for the composition of Deuteronomy came from Clements, "Deuteronomy," 165–166. For Yahweh as modeled on El (and also on Baal), see Wright, *Evolution of God*, 110–130. For Yahweh having a consort, see Wright, *Evolution of God*, 118–124. For an older argument that Jewish monotheism began with Moses, see Weinfeld, "Israelite Religion," 485–492 (which I also find interesting as a discussion of the differences between polytheism and Hebrew monolatry, which he calls monotheism). For specific arguments against Weinfeld's account of the origin of Jewish monotheism, see Halpern, "Monotheism," 526; Sperling, "God in the Hebrew Scriptures," 3–6; and especially Wright, *Evolution of God*, 131–164.

8. Sperling, "God in the Hebrew Scriptures," 4–5.

9. Judaism, Christianity, and Greek philosophy all share a similar idea of the soul as ultimately individual. In Hebrew as in Greek, all the various terms for "soul" (*nefesh* or *nepeš*, *neshamah* or *nishmah*, and *ruaḥ*) portray a human being as a discrete person. They usually refer to a property or activity of the body or to a living being as a whole. Both nefesh and neshamah refer to, among other things, the breath, the inner life-force of an individual organism. Other meanings also indicate a separate, distinct entity made of body, vital breath, appetites, emotions, intelligence, consciousness, and the like. It is the soul, for example, that suffers or rejoices according to changing circumstances. None of the terms imply a soul that is in substance God. *Ruaḥ*, which is often translated as "spirit," does connote something made or given by God, but in the sense of God as a distinct being creating or giving something to another being (Bemporad, "Soul: Jewish Concept," 450).

In spite of the soul designating the standpoint of the individual, Judaism maintained a strong sense of collective identity, with the soul relating to Yahweh principally through the Jewish community. The covenant between God and humans was with the Jewish community much more than with individual souls.

10. For example, Deut. 28:12, 15.

11. For example, Isa. 45: 5–7, 21–22. Dates for Isaiah come from Sawyer, "Isaiah," 328.

12. In Judaism, "the seeming contradiction between divine foreknowledge and human free choice . . . is not explicitly encountered in the Bible. . . . Exegetes grapple with the implications of these texts, and their proposals may be judged plausible or strained. What is important for us, however, is that the Bible itself does not address the issues" (Carmy and Shatz, "The Bible as a Source," 25–26). I find the contradictions between these God and human agencies even more extreme in the Qur'an than the Tanakh. Sûrah 35, for example, admonishes us in verse 5 not to be deceived by Satan: "O people, Allah's Promise is true, so do not let the present life delude you and do not let the Deceiver [Satan] delude you concerning Allah." Then the *same* sûrah tells us only three versus later that we have no choice in the matter: "Allah surely leads astray [whomever] He wishes and guides [whomever] He wishes . . ." (Qur'an, trans. Fakhry).

13. For God as king (*malek*) in the Tanakh, see Emerton, "Names of God," 549.

14. Nondualist accounts have their own problems in this area; the problems are just different. In Hinduism, the question is not whether we are free or fated or, perhaps, how we could be both; the question is rather, Why do we experience ourselves to be individual, mortal selves when we are Brahman, the universal, eternal self? Buddhism faces a different question: How can there be choice or virtue or enlightenment when there is no separate, enduring self to choose or to be virtuous or enlightened?

15. Plato develops his idea of soul (*psuchè*) within the context of the Greek meaning of the word. For the Greeks, a human was above all a living being, with that life evidenced primarily in breath animating a body composed of flesh. *Psuchè* thus meant breath but also the principle of individual life, as distinct from the body (*sōma*), which was what remained when life was gone (White, "Human Person," 295).

16. The "Good" is the archetype for all instances of good. Plato speaks of the "good" (e.g., a "good man") throughout *The Republic*, but perhaps best known is his analogy of the highest "Form of Good" to the sun and its illuminating power (508c–509b).

17. In the *Timaeus*, Plato characterizes the demiurge (creator) as much like a craftsman (28a–29). He also speaks of the demiurge as the creator of "order" out of chaos (30a).

18. I discuss Aristotle's substance in chapter 3.

19. Aristotle, *Metaphysics*, xii (trans. preface), 1071b3–1073a14, 1074b15–1075a11 (Λ: 6, 7, 9).

20. However, the Sophist Gorgias of Leontini (c. 483–376 B.C.E.) gave the first clear treatment of the philosophical problem of free will (Blackburn, *Oxford Dictionary of Philosophy*, 161).

21. Sedley, "Stoicism."

22. Per the Tanakh, God created living creatures "of every *kind*" (Gen. 1:20–21; my italics), while Adam gave to each creature (presumably, each *kind* of creature) their names (Gen. 2:19). For Maimonides, universals are merely ideas formed in our minds by the intellect, but our intellect accomplishes this task only to the extent it expresses God's own perfect intellect (Maimonides, *Guide for the Perplexed*, 1:72; 3:17,18 [Friedländer 118–119, 288, 289]).

23. Esp. chapter 1, note 9, and chapter 2, note 6.

24. Brainard, *Reality and Mystical Experience*, 210–213; McGinn, *Foundations of Mysticism*, 89–100; Quispel, "Gnosticism."

25. While Neoplatonic teachings were largely condemned by mainstream versions of Judaism, Christianity, and Islam, they had an enormous impact behind the scenes,

especially on the theology and mystical traditions of these religions. They influenced the entire Jewish mystical tradition, the Kabbalah, and many of its most important writers, such as Isaac Luria, were thoroughly Neoplatonic. Neoplatonism also influenced Jewish philosophers such as Isaac Israeli, Solomon Ibn Gabirol (Avicebron), Abraham Ibn Daud, and Maimonides. Christianity, which was less tolerant of Neoplatonic ideas than its two sister religions, still could not avoid its influence. Among well-known Christian mystics accused of being too Neoplatonic are Dionysus the Areopagite (Pseudo-Dionysus) and Meister Eckhart. Neoplatonism affected mainstream Christian theology in the works of the Cappadocian fathers (Basil of Caesarea, Gregory of Nazianzus, Gregory of Nyssa), Augustine, Aquinas, John Duns Scotus, and Eriugena. Plotinus's greater support in Islam was due in part to a historical accident whereby Plotinus's writings were mistakenly included among Aristotle's and were seen as Aristotle's works on theology. Neoplatonic ideas showed up in the philosophy of Mullā Ṣadrā, al-Kindī, and Ibn Sinā (Avicenna) and in the teachings of practically every Islamic mystic.

For an overview of Plotinus, see O'Brien, *Essential Plotinus*, 13–32 (for his discussion of Plotinus's "One": 17–21). Also see Brainard, *Reality and Mystical Experience*, 207–208.

26. Abrahamic mystical traditions typically talk of spiritual enlightenment differently from their Hindu counterparts. The Hindu worldview holds that union with God is typically a question of overcoming ignorance and wrongly directed desires and behavior. Within the Abrahamic worldview, however, union with God (to whatever extent it is possible) usually requires us to obliterate or transcend our mortal selves in some sense. Thus Islam speaks of *fanā*, or psychological "self-annihilation," and Christianity speaks of "ecstasy" (in the Latin sense of *ecstasis*, "standing outside oneself"). Unlike Hinduism, our difference from God is typically a *real* difference, not merely an appearance. Union with God requires first the destruction or transcendence of that reality wherein we are different from God. (See Louth, *Origins of the Christian Mystical Tradition*, 32–35, as well as his index entries on "ecstasy." Also see McGinn, "Love, Knowledge, and *Unio Mystica*," who discusses at length Christian ecstasy; again, see his index.)

27. Louth, *Origins of the Christian Mystical Tradition*, 75–77.

28. A few of the better-known Christians who spoke about some variety of mystical union with God are Bonaventure (c. 1217–1274), the anonymous author of the *Cloud of Unknowing*, Mechthild of Magdeburg (c. 1207–1282), Hadewijch of Brabant (unknown), Meister Eckhart (c. 1260–1328), Teresa of Avila (1515–1582), John of the Cross (1542–1591), and Giordano Bruno (1548–1600). See McGinn, "Love, Knowledge, and *Unio Mystica*," 59–86.

29. Islamic philosophy developed other interesting options as well, such as an atomistic alternative to the classical Aristotelian idea of substance. Perhaps adapted from Indian sources (Haq, "The Indian and Persian Background," 53–54), Islamic writers proposed that the world was made up of indivisible and indistinguishable primary particles, much like the *dhammas* in Buddhism. Unlike Buddhism, however, it was God that held the particles together over time or related them to each other in space rather than the mind or a natural process such as interdependent arising. This view continues to be discussed today (search: [Islam kalam atomism]). It is also a subject of debate in Western theology where God has been proposed as a solution to certain problems relating to causation in nature (the theory called "occasionalism").

30. Ibn-al'Arabī (1165–1240 C.E.), one of the best-known Abrahamic mystics and philosophers, writes, "If you insist only on His [Allah's] transcendence [to the cosmos], you restrict Him. And if you insist only on His immanence [within the cosmos] you limit

Him. If you maintain both aspects you are right. . . . You are not He and you are He and You see Him in the essences of things both boundless and limited" (Ibn-al'Arabī, *Bezels of Wisdom*, 75). (For more on Abrahamic mystical union with God, see Idel and McGinn, *Mystical Union*, and Sells, *Mystical Languages*.)

Chapter 9

1. Out-of-body experiences are just one, albeit striking, example of the mind's ability to generate a world that seems real and independent of ourselves. Dreams, illusions, and hallucinations (such as induced by drugs or lack of sleep) are some others.

2. For various theories and contemporary research regarding out-of-body experiences, see Blackmore, *Consciousness*, 355–361. Also, the chapter in which Blackmore covers out-of-body experiences, "Altered States of Consciousness," gives additional context for my discussion here.

3. Crane, "Problem of Perception," more fully discusses the arguments from illusion and hallucinations.

4. See, for example, Geertz, *Interpretation of Cultures*, 49.

5. Some researchers (e.g., Noam Chomsky, Claude Lévi-Strauss) have emphasized universal structures in language or society, while others, notably postmodern and poststructural philosophers (e.g., Jacques Derrida, Michel Foucault, Richard Rorty), have stressed the cultural relativity of language and ideas.

6. Following Franz Brentano (1838–1917), phenomenology has usually considered human awareness—"consciousness"—to be "intentional"; it is a "consciousness of" some object. Consciousness is thus thought of as encompassing both the *act* and the *object* of consciousness (Husserl's *noesis* and *noema*). My distinction between two roles of awareness concerns the *act*, or *process*, of awareness as opposed to its objects. The objects of awareness in its disjunctive and conjunctive roles are what I've called, respectively, "presence" and "publicity." (Publicity combined with the capability for presence is what I've called "everyday reality.")

7. My original intent in describing awareness's two roles as "disjunctive" and "conjunctive" was to suggest the logical "or" and the logical "and." However, I suspect that what I call the "disjunctive" role may be better thought of as a logical "not." First, the idea of "not"—of simple negation—seems to me to fit better with the idea of "pure consciousness" (see text box in this chapter on "'Intransitive' versus 'Transitive' Consciousness"). Second, both roles of awareness together then yield a logical "nand," which, when combined with other "nands," can express all the possibilities of Boolean logic. (For more on the universality of the logical "nand," search: [nand gate universal].)

8. For the activities of unconscious awareness, see chapter 5 section "Awareness and Conscious Awareness," especially note 17.

9. In considering the possibility of awareness having two roles, one of which is collective and unconscious, consider Benjamin Libet's 1985 finding that conscious choice is always preceded by roughly a half second of unconscious biological processes (Libet, "Unconscious Cerebral Initiative"). Consider also the 1992 discovery of "mirror neurons" that mimic certain activities in the brains of others. These neurons fire not only when an animal itself acts but also when the same action is observed in others; they also fire whether or not individuals are physically able to perform the actions. They have also been

associated with the development of social organization and language (Sanders, "Mirror System Gets an Assist"; Rizzolatti and Craighero, "Mirror-Neuron System").

    For a discussion of Libet's work and its philosophical implications, see Blackmore, *Consciousness*, 123–137. Also see the Wikipedia entry on the "Neuroscience of free will."

10.   Presence as content in awareness is paradoxical. If presence is itself content, then it should be something we can think about, refer to, or name (which, of course, I've done with the term "presence"). But to do so makes it a publicity, something that is sufficiently invariable for someone to name and refer to. And if this is the case, then it can't be what remains when all public properties are stripped away. (Recall the discussion of presence in the chapter 6 section "Enlightenment and 'Presence.'" See also the upcoming section "Sitting in a Café," as well as the discussion of the paradoxical nature of set theory in chapter 11.)

11.   See chapter 7 section "Individual and Collective Agency."

12.   When I say that our process of awareness "participates in generating a collective reality for us to experience," I'm not saying that our way of being aware creates matter, or ants, or trees, as they are independent of our awareness of them. I'm saying that our human way of being aware collectively generates the world as we perceive, understand, and influence it, as well as whatever else is uniquely human. It generates the properties of the universe that trace to our human way of being aware and that relate us effectively (or not) to properties that originate elsewhere.

13.   In this regard, notice the way that alternate accounts of the self align themselves with the two roles of awareness and reflect the tension between them. On the one hand is the picture of who we are that we get from Western monotheism (as well as from everyday experiences and much of Western philosophy). In this account, we are each individuals with natures based on the distinction between ourselves and what is other than ourselves. This account identifies us specifically with awareness's disjunctive role. On the other hand is the picture we get from certain Hindu and Buddhist philosophical schools (as well as from Descartes and the Enlightenment). In this account, everything we're personally aware of has its source in our minds, *including* our sense of self, the public world as it is present to us, and our everyday distinction between the two. Contrary to our everyday first-person experience, this picture identifies us with awareness in both its roles (with Hindus and Buddhists on different sides as to which role is more fundamental).

14.   The "capability for presence" denotes roughly what we often conventionally mean by "existence" (see chapter 7, note 6). I've pointed out that presence is a second-order property, a property descriptive of other properties. Existence is similarly a second-order property, and we can get some sense of presence as a second-order property by considering existence's role as such a property in formal logic. Formal logic associates existence with the notion of "number." To indicate that something exists (i.e., that something with certain properties has instances), formal logic uses the "existential quantifier" (symbolized by $\exists$). For example, $\exists x(Px)$ means "there exists an x such that x has the property P." Note that existence here is not itself a property of x in the way P is but rather a statement made about something having the property P (Blackburn, *Oxford Dictionary of Philosophy*, 129). Note, too, that if logic treats existence as a property of other properties, then these other properties cannot be used to describe existence, nor can existence be meaningfully used to describe itself.

    This second-order quality of existence has an important bearing on the very old "ontological argument" for the existence of God, which tries to use pure logic to deduce that God in fact exists and which seems to succeed because it doesn't treat existence as a second-order property. Saint Anselm of Canterbury (1033–1109) introduced the original version of the argument. He characterizes God as "that than which nothing greater can be conceived." If we agree that a being with existence is greater than one without existence, then God must exist, since if God didn't have existence, we could always conceive of a greater being that did have the quality of existence. Anselm's argument immediately had critics but has nevertheless been of considerable interest to philosophers (e.g., Leibniz used a version of it). Kant, Russell, and others have pointed out that "existence" can't be used in logical arguments in the way Anselm used it. Like "presence" in my use of it here, it is a second-order, not a first-order, property (Hick, *Philosophy of Religion*, 15–20). For Leibniz's version, see Look, "Gottfried Wilhelm Leibniz," §7.1.1.

15. Again, the term "awareness" here encompasses both conscious and unconscious awareness and behavior.

16. Van Gulick, "Consciousness," §2.1. Grim, *Philosophy of Mind*, 382–383.

17. See chapter 4 section "The Different Philosophical Schools."

18. For example, see Forman, *Problem of Pure Consciousness*; First Panchen Lama, *Great Seal of Voidness*, 20–27.

19. I focus here on what is often called the "hard problem" of consciousness. The distinction between the hard problem and the easier problems of consciousness comes from David Chalmers ("Facing Up to the Problem," §2). For other descriptions of this consciousness problem, see Güven Güzeldere's introduction to Block et al., *Nature of Consciousness*, 24; or Guttenplan's introductory essay to Guttenplan, *Philosophy of Mind*, 48, 82–86. For brief summaries of *views* on the subject written by their proponents, see the entry by Chalmers on "Consciousness" in *Oxford Companion to the Mind*, 205–218. For a much more thorough discussion, see Blackmore, *Consciousness*.

20. The views "from the inside" and "from the outside" are contrasted in Dennett, "Consciousness," 161.

21. Chalmers, "Consciousness," 207. Abundant online sources discuss awareness in terms of the distinction between first-person and third-person views (search: [consciousness first-person third-person]). See also Güzeldere's introduction to Block et al., *Nature of Consciousness*, 24; and Blackmore, *Consciousness*, 9.

22. Note that the consciousness problem, as Chalmers indicates, has nothing to do with an inability to explore empirically the mind side of this Cartesian divide. From research in psychology, we've learned a great deal about our first-person subjective life and have proposed theories to explain this data. Consciousness itself has been directly explored by phenomenologists, other philosophers, and psychologists, and theories have been proposed regarding its structure (e.g., higher-order theories of consciousness). This consciousness problem concerns instead the relationship between two kinds of data (that are *both* "third-person" in my context here), which I propose is best thought of in terms of how different sources of properties relate to each other.

23. For the mysteriousness of presence, see chapter 6 section "Enlightenment and 'Presence.'"

Chapter 10

1. As I pointed out in chapter 3, first-person and third-person views do not exactly follow their grammatical distinctions in my analysis of awareness. In grammar, the first-person

plural ("we") can be either distributive (i.e., individual) or collective. In my analysis, the first-person plural in the collective sense is a *third-person view*.

2. Generally speaking, the process by which objects come into consciousness is an unconscious one. This always seems true at the neurological level. However, there are important exceptions in cases of higher-level brain functions, such as if we are learning something and the matter of interest is difficult and needs to be worked through. For example, if I've memorized the sum of 7 and 5 as 12, I will add these numbers without conscious attention. But if I must focus to get the sum, then the process of arriving at 12 is not as automatic or unconscious. (For more on the subtleties of this distinction between conscious and unconscious awareness, see Baars, "Contrastive Phenomenology.")

3. Even if a property has an empty class, it still is something people can talk and think about.

4. In considering matter as its own COR, consider also Aristotle's "material cause"—the material out of which something is made as the cause of or explanation for why something is as it is.

   To review how human-specific properties are not in matter by itself, see chapter 6 section "Universals as Arising through Causal Interaction."

5. CORs can play different roles in defining properties. Some CORs are "primary" in that they produce publicities that are actually presented. Others are "secondary"; they produce publicities involved in background activities. For example, part of what makes a physical desk different from a condominium rule is the physicality of the desk. The desk's physical properties are part of what the desk is for me and presumably others; matter here is a primary COR. On the other hand, memorized condominium rules no doubt depend on physical brains for their existence, but I don't associate physical qualities with the rules *themselves*. The rules present themselves as ideas that come from and are understood by people, not matter.

   In the case of the condominium rules, note that the distinction between "primary" and "secondary" CORs does not distinguish between those who determine the rules (our board of directors) and those to whom the rules apply (the members of the association). How rules and other properties become public within groups of people is a complex, extensively discussed subject of sociology, psychology, politics, and law, and it is outside the scope of this book.

6. Georges Rey (in "Question about Consciousness") discusses numerous theories used to explain mental phenomena and notes that all of them can be fairly easily explained *without* any need for consciousness. In other words, so far we've found no defining properties of consciousness that couldn't at least theoretically be mimicked by a machine.

7. Two traits often used to distinguish living beings from matter are metabolism and reproduction. In the context of this book's terminology, both these traits relate living beings to CORs. Metabolism pertains to the *individual members* of CORs. It describes the processes that maintain an individual's ability to participate in a COR and instantiate nonphysical properties. Reproduction, on the other hand, doesn't describe the maintenance of one COR member over time; rather, it describes the maintenance of *CORs themselves* in space and time—the production of other spatially distinct individuals with similar ways of being aware that ensures the continuation of CORs over generations.

8. Nagel, "What Is It like to Be a Bat?"

9. In this context, consider John Searle's "Chinese Room" argument against human minds being merely computers (e.g., Blackmore, *Consciousness*, 204–206; or see online for

more information on the "Chinese Room" argument). In my example, a cell-like robot would be similar to Searle within the Chinese room. Like Searle translating Chinese without knowing what the Chinese characters meant, the robot cell would respond to light and darkness without knowing what light and darkness would mean to it were it a member of a COR that originated/maintained the meaning.

10. For an examination of cellular life and how it is much more similar to human life than we might think, see Niehoff, *Language of Life*. Notice her emphasis on language (i.e., intercellular communication).

11. In the context of how individuals help maintain groups, consider how natural selection would seem to make sense only to the extent that the behavior of individuals furthers the classes of beings to which the individuals belong.

Consider first how evolution does not require individuals themselves to survive; like any particular, they have finite (though for many species, almost indefinite) life-spans. Instead, it is the group with its genome that is selected and survives. In this respect, the important issue in evolution and species characteristics seems to be not the individual's nature per se but rather the service of the individual's nature to the group as a whole (Brooks, *13 Things*, 122ff., describes the controversy over individual versus group selection).

Second, consider how difficult it is to explain sexual reproduction apart from a group selection theory. Why do so many creatures reproduce sexually rather than asexually when sexual reproduction seems to come at such a cost of time, energy, and specialized biological mechanisms? If only individuals were selected, then why wouldn't an individual choose to clone itself (Brooks, *13 Things*, 136ff., 144)?

For these and other reasons (e.g., certain animals' natural altruistic behavior), many biologists favor a "group" selection theory; however, the mechanism for this process remains difficult to understand. (See also Sewall Wright's works; Lloyd, "Units and Levels of Selection"; Price, "The Controversy"; Berreby, "Enthralling or Exasperating.")

12. Ludwig Wittgenstein speaks of words as similar to tools we create whose functions vary with context: "Think of the tools in a tool-box: there is a hammer, pliers, a saw, a screwdriver, a rule, a glue-pot, glue, nails and screws.—The functions of words are as diverse as the functions of these objects. (And in both cases there are similarities.)" (Wittgenstein, *Philosophical Investigations*, 6e:11).

13. Physicists model quantum phenomena using quantum field theory, which emphasizes the collective over the individual. It describes matter as being made up of "fields," which give the quantum mechanical properties at all space-time locations; particles are excitations of fields. Nevertheless, what we observe when we examine these fields closely enough are always particles. (The individual and collective in matter are further discussed in the next chapter.)

14. Panpsychism (the view that matter possesses certain basic qualities of the mind) was the dominant philosophical strategy for reconciling mind with matter through the nineteenth into the mid-twentieth century, when it began to be supplanted by "emergentism" (the view that mental properties arise from but are not themselves found in matter). Recently, there has been renewed interest in panpsychic solutions to the mind-matter problem largely inspired by the work of Galen Strawson (Seager and Allen-Hermanson, "Panpsychism," §1, §3). (See also Holt, "Mind of a Rock," 19–20; Kafatos and Nadeau, *Conscious Universe*. For a TED talk by David Chalmers arguing the merits of panpsychism, see Chalmers, "How Do You Explain Consciousness?")

15. Consider here Kant's "forms" of space and time in his *Critique of Pure Reason* (A19, B33–A49, B73 [Smith 65–91]).

Chapter 11

1. Seager and Allen-Hermanson, "Panpsychism," §1.
2. Per Aristotle, "What causes each of them [a particular thing] to be one? For in anything which has many parts and whose totality is not just a heap but is some whole besides just the parts, there is some cause [of the unity]" (Aristotle, *Metaphysics*, 1045a 9–10). John Stuart Mill began the modern debate on the subject of emergence with his discussion of causality in his 1843 book *A System of Logic, Ratiocinative and Inductive* (bk. 3, ch. 6).
3. See the chapter 3 text box "The Different Compositions of Particulars and Universals."
4. Weak or epistemological emergence can be defined in terms of what an observer is able to know of a complex system. As such, it pertains principally to language and to the concepts we use to understand our world (O'Connor and Wong, "Emergent Properties," §2–3).

    For a fairly recent proposal for an emergent theory of mind, see Vision, *Re-emergence*. For a criticism of Vision, see Goff, "Gerald Vision: *Re-emergence*." For a discussion of emergent theory as it might relate to religion and the concept of God, see Clayton and Davies, *Re-emergence of Emergence*, which has a section on this subject.
5. Note here how the part-whole construction of particulars also exhibits a set-member relationship, the class logic that distinguishes universals.
6. With our conventional association of awareness with human-like sentience, it may seem that "awareness" is not the best word to use for such a rudimentary source of properties as matter. One might agree to an underlying reality-generating process at work throughout the cosmos, but why call that process "awareness"?

    I find the word apt because it seems to me that it is precisely this reality-generating process that we call awareness in ourselves. The things that specifically humans create—our word meanings, laws, money, food recipes, works of art, and clocks as ways to tell time, as well as our subjective experiences of love, pain, hope, desire, and the rest—wouldn't exist without our human awareness. I've expanded the term's use to include matter because this reality-generating capability appears here as well, albeit with very different products. (Cf. Alfred North Whitehead's use of the word "prehension" in Whitehead, *Process and Reality*, 18–20, 22, 23, 217ff.)
7. "Network theory" is another proposal to trace properties to the collective structure of relationships as opposed to particular things (Goldman, *Science in the Twentieth Century*, 2:20–21).
8. Niehoff, *Language of Life*, 21.
9. For more on scale as a dimension of reality, see postscript 1 on this subject.
10. Einstein, "Geometry and Experience," §2. In the paragraph following the one quoted, Einstein gives his own answer: "As far as the laws of mathematics refer to reality, they are not certain; and as far as they are certain, they do not refer to reality." The extraordinarily accurate values given by quantum mechanics since then would seem at odds with his conclusion.
11. Johnson, "String of Numbers," 1.
12. Ibid.
13. The quote from Wigner is the title of a lecture he gave at New York University in 1959 (Wigner, "Unreasonable Effectiveness of Mathematics").
14. Maddy, "Set Theory"; Jech, "Set Theory," 1. The modern, mathematically rigorous version began with Georg Cantor (1845–1918).

15. "Modal" logics include operators ("modals") that are not derived from set theory but rather qualify logic's set-member relationships in fundamental ways. In a narrow sense, modal logic deals with "necessity" versus "possibility" (as distinguished from "probability," which is mathematically based and thus can be derived from set theory). In a broader sense, modal logic includes "deontic logic," which deals with morality (obligation and permission) as well as logics concerning time and conditional statements (e.g., If x had happened, then y would have happened).

These kinds of logical operators raise the question of where the source is for these modal qualities. If we can presume that awareness is, in some sense, that source, then another question arises. What activities of awareness do these logical operators point to that is not explained by awareness's publicity-producing behavior (which is inherently set theoretical)?

16. Maddy, "Set Theory"; Jech, "Set Theory," 1.

17. "Naive" set theory is distinguished from "axiomatic" set theory, which has restrictions to remove the paradoxes of naive set theory and make it more mathematically useful.

18. Russell's paradox and others are easily found online (search: [set theory paradoxes]). Apart from online sources, I used, in part, Maddy, "Set Theory"; O'Connor and Robertson, "History of Set Theory"; Hofstadter, *Gödel, Escher, Bach*, 19–21; Rundle, "Logic: History of," 556.

19. The indeterminacy demonstrated by the first of Gödel's two indeterminacy theorems is similar to the liar's paradox, but with "true" replaced by "provable." The liar's paradox is the sentence "This statement is false" (or a variation thereof). If the statement is true, then it is false; if the statement is false, then it is true. What Gödel says is that for every formal, nontrivial system of arithmetic (i.e., of natural numbers: 0, 1, 2, 3, . . .), one can generate a true statement (one generated in accordance with the system's rules) that is neither provable nor disprovable in that system. Gödel's second indeterminacy theorem expands this notion to show that such a system can never demonstrate its own logical consistency (Hofstadter, *Gödel, Escher, Bach*, 15–19; Raatikainen, "Gödel's Incompleteness Theorems").

20. The history of theories of light comes principally from Holton and Brush, *Introduction to Concepts*, 384–394, 427–445; and Gribbin, In *Search of Schrödinger's Cat*, 7–100.

21. Many metals emit electrons when light shines on them. This "photoelectric effect," first discovered by Heinrich Hertz in 1887, led to several problems. Particularly troublesome was that, regardless of how bright the light, the electron discharge stopped completely when experimenters decreased the light's wave frequency below a certain threshold. Increasing the wave frequency of the light does increase the light's energy, but why did specifically the *frequency* as opposed to the *brightness* (i.e., quantity) of light have to reach a certain energy level before anything happened? Einstein's 1905 paper, which later won him a Nobel Prize, proposed that this phenomenon could be explained if light came in tiny bundles whose energy depended on frequency. Regardless of how many bundles hit the plate at the same time (i.e., regardless of how *bright* the light), if the individual bundles did not have enough energy in themselves (a high enough frequency), they couldn't kick any electrons off the plate.

22. Von Baeyer, "World on a String," 13. Reprinted with permission from John Wiley and Sons, Hoboken, NJ.

23. See, e.g., Moskowitz, "Tangled Up in Spacetime."

24. Physicists will claim that what really exists are only waves as defined by quantum field theory; particles are simply excitations of fields (Carroll, *Higgs Boson and Beyond*, 34).

No doubt this is correct in that what exists of quantum phenomena apart from observation are always waves. But note that such a claim depends on our Western concept of reality as that which is mind independent (or, in this case, that which is observer-independent), and it doesn't seem to explain why there is also an equally viable Heisenberg's picture of quantum phenomena, a picture that begins with the quanta of quantum mechanics rather than with their collective, probabilistic behavior.

25. Quoted in Gribbin, *In Search of Schrödinger's Cat*, 164.

26. Notice in this paragraph and elsewhere how everyday reality as defined here can be rendered in terms of the information (publicity) that is available (capable of presence) in a system (a COR). Framing physical reality in terms of information has been helpful for understanding wave-particle duality as well as other physical phenomena. For example, researcher Stephanie Wehner reports that it reveals a close connection between wave-particle duality and Heisenberg's uncertainty principle, a connection that "comes out very naturally when you consider them as questions about what information you can gain about a system" (Wehner, "Two Quantum Mysteries Merged into One," §3).

   Also of interest in the context of this book's terminology is how this uncertainty principle describes the limit for how accurately physical phenomena can be measured: it describes this limit in terms of certain *pairs* of measurements. The two most often mentioned measurements are "position" and "momentum." The position of an object or event points to its presence at (or spread over) a *particular* space and time. Momentum describes an object or event in terms of velocity and thus points to a *transition* between particular space-times—in other words, to *more than one* presence.

27. Sabine Hossenfelder writes, "The quantum behavior of macroscopic objects decays very rapidly. This 'decoherence' is due to constant interactions with the environment which are, in relatively warm and dense places like those necessary for life, impossible to avoid. This explains that what we think of as a measurement doesn't require a human; simply interacting with the environment counts" (Hossenfelder, "10 Quantum Truths," §10).

28. In our everyday lives, reasonable conclusions about the world we live in rarely fall into neat true-false categories. The inadequacy of conventional two-valued logic is evident from the many alternatives one can find in an internet search: "modal" logic, "second" and "higher-order" logic, "many-sorted" logic, "quantum" logic, "intuitionistic" logic, "three-valued" logic, "fuzzy" logic, "free" logic, "relevance" logic, "linear" logic, and "non-monotonic" logic. The inadequacy of two-valued logic is also evident in Indian logic, which historically emphasized real-life, practical circumstances and long ago developed types of logic that have more options than just true and false (Brainard, "Defining 'Mystical Experience,'" 386).

29. Understanding reality as a fugue requires understanding reality as originating with CORs and keeping the CORs straight when talking about what is real.

   Suppose you have a very elderly aunt to whom you have not spoken in awhile, and you have no idea whether she is still alive. Insofar as whether she is alive or dead, what is the reality here?

   The answer depends on what reality you are talking about. From your perspective, you don't know if she is alive or dead; that's the reality you'd act on if, for example, you wrote to or phoned her. On the other hand, your aunt as an enduring being is the collectivity of reference for all properties originating with her; it is to her reality that we would typically defer to answer such a question if at all possible. Note that what is real is not a question of your aunt's reality being "right" and your reality being "wrong." *Both* realities make up this universe we live in and play their parts in determining our choices and interactions.

30. In regards to this superposition of CORs, consider also the dimensions of time and space and how *normal* it is in our own everyday experience for an event in one COR to occur at a different time and place for another COR.

Consider a bag of groceries that I've purchased and put in the trunk of my car. On the way home, the bag falls over, but I'm not aware of this happening until I get home and open the trunk. If I then were asked when the bag fell over, I'd say sometime on the drive home, because that's when the *physical* bag fell over. If a jar of spaghetti sauce broke, then that's when the chemical reactions would start that would soil my car and cause an odor. Nevertheless, that's *not* when the grocery bag expressed the property of having fallen over insofar as *I* or any other *human* was concerned. It was not until I opened the trunk that the bag became *for me* a mess on the floor. *Up to that time, all my behavioral interactions affecting the universe reflected a different state of affairs.* This is also the case for spatial location. The physical grocery bag may have fallen over someplace on the way home, but the fallen bag impacted me in my driveway at home.

31. Many physicists today find reality best thought of in terms of many parallel universes, each expressing an alternate possible outcome for the collapsing probability waves of all the quantum mechanical interactions in our universe (the "many-worlds interpretation" of quantum mechanics). This book's concept of reality also provides for innumerable, alternate paths for reality; however, it differs in that all these possible worlds define *our own* universe, the one we ourselves live in. (See also the chapter 7 text box "Counterfactual Conditionals." For more on this idea of "universally true," see postscript 2, "A Definition for Truth.")

Chapter 12

1. Kant is an important, influential example of a Western philosopher who, in *Critique of Pure Reason*, does speak of the limits of reason and of ultimate reality as beyond what reason is capable of understanding.

2. I expect one could easily find Buddhists who claimed that their account of reality was very logical and could be understood by a clear-minded analysis of one's experience. There are grounds for this. As discussed earlier, Buddhism begins with our own everyday experience and analyzes it as rationally as possible (in this respect, it resembles modern empirical science). Nevertheless, what is sometimes overlooked by such a claim is the difficulty Buddhist philosophers themselves have had reconciling the Buddha's no-self doctrine with the concepts of virtue, karma, and enlightenment or explaining universals in a cosmos marked by impermanence (chapter 6, note 13). Then, too, Nāgārjuna himself distinguished between two truths: *saṁvṛti satya*, which is empirical or conventional truth, and *paramārtha satya*, the highest truth, which concerns what is beyond the conventional. Moreover, as I also discussed earlier (chapter 1 section "Science's Limitations"), the philosophical foundation for empirical science is itself problematic.

All this is to say that nirvana in Buddhism is beyond conventional logic and reason, which, as I also noted in chapter 6, even the Buddha himself acknowledged.

This contrast between empirical truth and the highest truth is more explicit in Nondual Hinduism, although here the distinction comes as one between levels of being or reality, of which the second highest is empirical reality (chapter 4, note 18). (See also Brainard, *Reality and Mystical Experience*, 111–115, 165–167.)

3. See chapter 6's discussion of interdependent arising and also Nāgārjuna (with related text box and notes).

4. When I say religions put awareness at the center of their concepts of reality, I'm referring to some variation of awareness as defined in this book, not specifically human "consciousness" (not specifically *vijñāna*). As discussed earlier, we tend to use the word "consciousness" to refer to a *type* of awareness, the way we humans are aware when we're awake and attentive. Awareness as I use it intends to signify that which in Buddhism distinguishes a creature as "sentient" (and which I've argued also appears to underlie matter). Awareness is that presencing-of-publicity process which gives birth in every moment to the universe—although the "moment" of birth might be as short as a quantum instant or as long as perhaps the universe is old and still incomplete. (See also postscript 1, "Scale as a Dimension of Reality.")

5. The Sphinx asks Oedipus, "What walks on four legs in the morning, two in the afternoon, and three at night?" Oedipus answers that it is ourselves. We crawl on four legs when young, walk on two when older, and use a cane as a third leg when older still.

6. See chapter 3 section "The Ambiguous Relationship of Universals and Particulars" and chapter 6 section "Enlightenment and 'Presence.'"

7. A universal differs from a particular according to a publicity's spatial scope (see chapter 7, note 22). The distinction between universals and particulars seems to me to have its source in the different way occasions of awareness group together in space versus time (see chapter 10 section "Our Concept of Self in Space and Time").

8. See chapter 3 section "Some Historical Roots of the Universal-Particular Distinction" as well as the different Hindu and Buddhist views on this subject.

9. See chapter 7 section "Individual and Collective Agency."

10. See chapter 11 section "Two Pictures of Physical Reality."

11. See chapter 11 section "Mathematics and the Behavior of the Universe."

12. See chapter 1 text box "Logical Positivism and the Limits of Science." Greek philosophers like Plato presumed that aware beings—souls—related universals to particulars (see chapter 8 section "Philosophical Influences").

13. William James makes a similar observation as to the inexplicable nature of will: "We know what it is to get out of bed on a freezing morning in a room without a fire, and how the very vital principle within us protest against the ordeal . . . now how do we *ever* get up under such circumstances? If I may generalize from my own experience, we more often than not get up without any struggle or decision at all. We suddenly find that we *have* got up" (quoted in Brooks, *13 Things*, 160).

14. Ludwig Wittgenstein says the following about the reality (the "life") of words or concepts: "Every sign [e.g., word, symbol] *by itself* seems dead. *What* gives it life?—In use it is *alive*" (Wittgenstein, *Philosophical Investigations*, 128e:432).

15. Martin Heidegger writes that the "Being [of ourselves in the world] reveals itself as *care*" (*Being and Time*, 182 [Macquarrie and Robinson 227]). "Care" is a central concept in his *Being and Time*; we are born into an inherently oppressive existence and propelled toward overcoming the anxiety caused by this predicament (ibid., 180–181 [225–226]).

16. In regards to my associating intention and value with the origin of reality, consider how Whitehead puts creativity and novelty at the root of all reality (Whitehead, *Process and Reality*, 7, 20, 21–21).

17. See chapter 4 section "The Five Cloaks of Brahman."

18. To appreciate the distinctive nature of values as properties, consider first how values are second-order rather than first-order qualities. They dispose us to behave in one way or another *in relation to* the first-order properties by which we recognize objects or events. We can't know how we'll feel about something without first distinguishing what it is—without first knowing whether it's a meal for us or we're a meal for it. The second-order status of values comes out clearly if we consider Locke's distinction between primary and secondary qualities (Locke, *Essay concerning Human Understanding*, II:viii). If we accept his distinction, then the values we ascribe to objects are properties of properties—that is, second-order properties.

Consider also how we explain nonhuman behavior differently than human behavior. According to the dominant theory, we typically explain the behavior of nonhuman phenomena by pointing to a "covering law"—a rule of nature that the particular event instantiates. For example, we might explain why a car skidded by pointing to icy road conditions. Since ice is by nature slippery, this quality of the road explains why the tires slipped in this instance. On the other hand, what we want when explaining the behavior of human beings is typically not a covering law but rather the person's state of mind, the *values* that prompt his or her behavior. For example, we might explain why a person ate a sandwich by saying that he was hungry. Note here how explanations of our aware behavior seem to have a built-in teleology; the behavior aims at a future valued result. To act out of hunger is to act in order to allay the hunger (principally from Kim, "Explanation").

Also consider Aristotle's "final cause"—the goal or purpose of an action as a cause of the action. Consider as well Thomas Nagel's *Mind and Cosmos*, at the center of which is an argument for the importance of a "natural teleology" of some form in any complete description of our universe.

19. While life may have joys as well as sorrows, pleasure as well as pain, the point made by the term *samsara* is that life conceived of from the standpoint of individual, mortal existence is *fundamentally* unsatisfying.

20. Both the title of this section and its opening quote come from the Chinese Zen Buddhist teacher Wu-men Hui-k'ai, better known by his Japanese name, Mumon Ekai (1183–1260 C.E.). Wu-men titles his compilation of koans "Gateless Gate" (in one translation), which strikes me as a good name for any means used to get past this individual-collective puzzle and our sense of self as a seemingly enduring individual.

The translation is from Reps, *Zen Flesh, Zen Bones*, 88, with Tuttle Publishing's permission to reprint and slightly modernize the translation.

21. There has been extensive study of the effects of meditation on the brain using brain imaging. On the basis of his research, Andrew Newberg argues that one of the brain's primary functions is "self-transcendence" (Newberg, "Are We 'Hard-Wired' for God?").

22. See chapter 4, note 13.

23. Recall that some Buddhist schools (e.g., Pure Land Buddhism) don't aim toward enlightenment in this lifetime.

24. Smith, *World's Religions*, 29–32.

25. Hart, *Art of Living*, 119–120.

26. Addiss, *Zen Sourcebook*, 89. A similar description of a koan's intent is given by Reps, *Zen Flesh, Zen Bones*, 87.

27. Mystical traditions in the Abrahamic religions are often spoken of as involving, first, a *cataphatic* phase in which one studies monotheistic philosophy sufficiently to understand God in positive terms. This is followed by an *apophatic* phase in which one deconstructs, so to speak, that understanding in order to discover a higher truth. (Gregory of

Nyssa's path from light to darkness mentioned in the text is an example.) See Michael Sells, *Mystical Languages of Unsaying*.

28. Brainard, *Reality and Mystical Experience*, 222–225; or Louth, *Origins of the Christian Mystical Tradition*, 94.

29. See chapter 8 section "Philosophical Influences," especially note 26.

30. Note how spiritual practices aimed at a greater understanding of reality can impact our understanding of death. We all eventually die, but how we frame our eventual death depends in part on the extent to which we identify ourselves with our physical bodies and their individual points of view. All the religions discussed here argue that identification with our physical bodies and their perspectives is in some sense a consequence of false beliefs about ourselves and/or misguided attachments. They also argue that such beliefs or attachments can be overcome. (I've also argued that, in many ways, we already do not identify ourselves with our physical bodies and their individual points of view. We see the world out of collective eyes and think and feel with a collective brain. I've also claimed that our bodily point of view stripped of its collective nature is nothing but a self-less presence—the point made by the Buddhist concept of *anatta*.)

31. All the practices mentioned here can be found online. Other possible searches are [mystical experience], [Kabbalah], [Sufism], [contemplative prayer], [samadhi]. See also Brainard, *Reality and Mystical Experience*.

32. Recall also the role of gods in building community in polytheistic cultures (chapter 8).

33. The movie *A Clockwork Orange* comes to mind, although in that case the mechanism to compel virtue was intensive psychological conditioning. Note also in this regard the Darwinian side to ethical behavior. Insofar as such behavior furthers group survival, we might expect it to be traceable, at least in part, to genes rather than conscious choice.

34. See chapter 8 text box "Free Will, Evil, and Accounts of Reality."

35. Brainard, *Reality and Mystical Experience*, 260–262.

36. Morreall, "Philosophy of Humor," §4.

37. For more on humor as incongruity and play, see Morreall, "Philosophy of Humor," §§4–6; or Smuts, "Humor," §2:c–d.

38. Consider in this regard Douglas Hofstadter's many examples of pleasurable puzzles in his book *Gödel, Escher, Bach: An Eternal Golden Braid*.

    Moving a little past incongruity theory, consider the social element involved in humor—how laughter at a joke often increases in strength as the humor gathers individuals into a collective feeling. Note, too, the social component to tickling; we don't seem able to tickle ourselves. In other words, humor also seems to be about an individual's relationship to others and to the community, which brings us directly back to the individual-collective puzzle, but now with an everyday enjoyable dimension.

    Śaṅkara, the foremost spokesperson for Nondual Hinduism, explicitly speaks of the puzzle that hides our oneness with the divine—maya—as the "sport" or "play" (lila) of God (search: [sankara maya lila]).

    For more on the social side of humor, see the "play" theory in the two articles mentioned in the prior endnote. For an "Enlightenment" theory of humor, see Karlen, "Humor and Enlightenment."

39. James, *Varieties of Religious Experience*, 387–388.

40. What lies on the other side of the gateless gates can seem extremely paradoxical. In speaking about our relationship to our religious symbols, Buddhism has a saying that "painted cakes don't satisfy hunger." In other words, religious symbols are just labels; they're not the reality they point to. However, the Zen Buddhist patriarch Dōgen bends

this saying back on itself by showing how *all* of our reality is, in itself, nothing but cakes painted by our minds and that it's through the "eating," so to speak, of the "painted cakes" of spiritual symbols that we arrive at the satisfaction of our spiritual hunger. In other words, painted cakes both do and do not satisfy spiritual hunger (Dōgen, *Flowers of Emptiness*, chap. 24 [Gabyō]).

Postscript 1

1. For more on space and time, see chapter 5, note 10; and chapter 10 section "Our Concept of 'Self' in Space and Time." Also see Sklar, *Philosophy of Physics*, 15–18. For other difficulties with our concept of space-time, see ibid., 15ff., and Nagel, *View from Nowhere*, 59.

2. Another classic illustration of what scale travelers might see is Kees Boeke's *Cosmic View: The Universe in 40 Jumps*.

3. Scaling in space-time rather than just space would also impact measurements of other physical properties. An object we at our scale observe to be steadily accelerating would appear to an observer at one-quarter scale to be accelerating four times slower. Assuming that Newton's second law of motion holds at both these scales and force remains equals to mass times acceleration, then the ratio of force to mass would also appear to be four times less when observed at one-quarter scale. At this scale, one would measure the mass of the accelerating object to be four times more or one would measure the force to be four times less depending on which was assumed to be invariable over the change in scale.

4. Slobodchikoff, *Chasing Doctor Dolittle*, 152.

5. Whitfield, "All Creatures Great and Small," 342.

6. While no writer I have read explicitly refers to the fourth scaling dimension as "time," all the various explanations I've seen for this extra dimension implicate time indirectly. More precisely, they implicate *energy* (metabolism, for example, is energy expended per unit of time). And energy, by definition, is measured in units that combine both space and time.

   For more on scaling laws, see Harrison, "Basal Metabolic Rate"; Klarreich, "Life on the Scales"; West and Brown, "Life's Universal Scaling Laws." Also see Milius, "Predator-Prey Relationship," 10. For research that specifically links perception of temporal information to metabolic rate and body size, see Healy et al., "Metabolic Rate and Body Size."

7. The question to ask for determining scale is, at what scale do the publicities—the properties, rules, and so on—originate that describe a thing's interactive behavior? Newtonian physics qualified by general relativity seems to work fine at all scales until we approach very small scales where the individual-collective dynamics of material particles start to produce significant uncertainties and infinities. In my context here, this suggests that the physical properties described by Newtonian physics and general relativity are generated at these smaller scales.

8. Specifically, the entropy of a system is Boltzmann's constant times the logarithm of the total number of microstates that will produce a certain macrostate: $S = k \log W$ (Carroll, *Mysteries of Modern Physics: Time*, 178).

9. Sean Carroll explains that "the way we coarse-grain the real world [i.e., define a macrostate] depends on what is actually easy for us to measure" (Carroll, *Mysteries of Modern Physics: Time*, 183). It is easy for us to measure temperature and pressure; it is very hard if not impossible for us to measure the position and velocity of every atom in a container of gas. Such an answer seems to me to make entropy a function of the scale of the observer.

If this is not Carroll's intent, then such an answer begs the question of exactly what it is that privileges a certain scale for defining a macrostate. Carroll writes, "The way that we coarse-grain the real world is very natural given how the laws of physics work and how *we actually measure things* in reality" (ibid.; my italics).

10. See chapter 3, note 3.

11. At least one physical theorist, Laurent Nottale, has attempted a theory of "scale relativity" based on "fractal space-time" (although he appears not to have answered his critics very successfully). As far as I can tell, Nottale associates a scale state with the resolution of measurement at that scale (Nottale, *Fractal Space-Time*, 2), which seems to implicate the scale of the observer or the observation. It's unclear to me, however, how similar his concept of observer might be to the concept of awareness I've proposed here.

Postscript 2

1. Aristotle, *Metaphysics*, 1011b25.

2. For example, Blackburn, *Oxford Dictionary of Philosophy*, 84–85, 323.

3. The difference I describe here between the ways we use "real" and "true" shows up also in the way logic's "existential quantifier," "∃," is used (see chapter 9, note 14).

4. It seems to me that the question of whether Monopoly is played with real money asks if Monopoly money is capable of being present in our collective lives in the same way that money usually is. In other words, can we use Monopoly money in the way we conventionally use money in our everyday lives—to buy groceries, for instance, or pay bills?

5. In this context, note how we use the word "real" to identify and emphasize CORs. Suppose I said that "Paris is a real city, but the Emerald City of Oz is not." What quality does the word "real" add to "city" in this statement? At least part of what "real" conveys here for me, and I expect others, is Paris's physical existence. Paris is not a city of just human imagination; it has a material presence, a physical COR. Suppose someone then rejoined by saying, "Nonetheless, the Emerald City is surely a real city in the world of our imaginations for anyone familiar with the story of the Wizard of Oz." In this case, "real city" conveys something different. The statement emphasizes the importance of nonphysical realities in our lives, the importance of human CORs irrespective of other CORs.

   We can also consider these statements in another way. If in a certain context the word "real" implies having a physical COR, then the Emerald City of Oz has no instances; it's not capable of presence. If "real" does not imply a physical COR, then the Emerald City of Oz does have instances; it is capable of presence.

6. Correspondence is not all or nothing; truth allows for a certain degree of *vagueness* in this correspondence. For example, I perceive a car before me and believe it to be a real car; I believe the publicities that are present to me correspond to publicities in fact produced by the human and material CORs of the car. But the truth of my beliefs about the car depends very much on the context. If I am dodging a car on the street, it makes no difference if the car is powered by gasoline or diesel fuel. If I plan to refuel, I need to have that detail correct. The important question here is whether the properties (publicities) I personally ascribe are close enough to those of the collectively maintained reality to suffice for the present occasion. In other words, our take on reality includes a judgment of sufficiency, which allows the correspondence to be somewhat vague without being false. (See also chapter 11, note 28. For a discussion of vagueness in logic, see Kahane,

Logic and Philosophy, 272. I discuss vagueness more thoroughly in Brainard, "Defining 'Mystical Experience,'" 366–369.)

7. For another example, see chapter 11, note 29.

8. For example, see Horwich, "Truth," for an overview of various theories. Individual theories are easily found online.

Bibliography

Addiss, Stephen, ed., with Stanley Lombardo and Judith Roitman. *Zen Sourcebook: Traditional Documents from China, Korea, and Japan*. Indianapolis: Hackett, 2008.

Aquinas, Thomas. *Summa Theologica*. In *Basic Writings of Saint Thomas Aquinas*, edited and annotated, with an introduction by Anton C. Pegis, translated by Laurence Shapcote (English Dominican Translation). New York: Random House, 1945.

Aristotle. *Aristotle's Metaphysics*. Translated with commentaries and glossary by Hippocrates G. Apostle. Grinnell, IA: Peripatetic Press, 1979.

———. *Aristotle's Nicomachean Ethics*. Translated with commentaries and glossary by Hippocrates G. Apostle. Grinnell, IA: Peripatetic Press, 1984.

Armstrong, Karen. *A History of God*. New York: Alfred A. Knopf, 1994.

Audi, Robert. *Epistemology: A Contemporary Introduction to the Theory of Knowledge*. New York: Routledge, 1998.

Baars, Bernard J. "Contrastive Phenomenology: A Thoroughly Empirical Approach to Consciousness." In *The Nature of Consciousness: Philosophical Debates*. Edited by Ned Block, Owen Flanagan, and Güven Güzeldere, 187–201. Cambridge, MA: MIT Press, 1997.

Ball, Philip. "The Strange Inevitability of Evolution: Good Solutions to Biology's Problems Are Astonishingly Plentiful." *Nautilus* 20 (January 8. 2015). http://nautil.us/issue/20/creativity/the-strange-inevitability-of-evolution.

Barnes, Jonathan. *Early Greek Philosophy*. New York: Penguin, 1987.

Barrow, John D. *New Theories of Everything: The Quest for Ultimate Explanation*. New York: Oxford University Press, 2007.

Bealer, George. "Property." In *The Cambridge Dictionary of Philosophy*. 2nd ed. Edited by Robert Audi. New York: Cambridge University Press, 1999.

Bemporad, Jack. "Soul: Jewish Concept." *The Encyclopedia of Religion*. Ed. in chief, Mircea Eliade. New York: Macmillan, 1987.

Berkeley, George. *Three Dialogues between Hylas and Philonous*. In *Classics of Western Philosophy: Second Edition*, edited by Steven M. Cahn. Indianapolis: Hackett, 1985.

———. *A Treatise concerning the Principles of Human Knowledge*. In *Classics of Western Philosophy: Second Edition*. Edited by Steven M. Cahn. Indianapolis: Hackett, 1985.

Berreby, David. "Enthralling or Exasperating: Select One." *New York Times*, September 24, 1996, Science Times section, C1.

Bird, Alexander, and Emma Tobin, "Natural Kinds." In *The Stanford Encyclopedia of Philosophy* (Winter 2016 Edition), edited by Edward N. Zalta. Stanford University, 1997–. Article revised January 27, 2015. http://plato.stanford.edu/archives/win2016/entries/natural -kinds/.

Bisiach, Edoardo. "Understanding Consciousness: Clues from Unilateral Neglect and Related Disorders." In *The Nature of Consciousness: Philosophical Debates*. Edited by Ned Block, Owen Flanagan, and Güven Güzeldere, 237–253. Cambridge, MA: MIT Press, 1997.

Black, Jeremy, and Anthony Green. *Gods, Demons, and Symbols of Ancient Mesopotamia: An Illustrated Dictionary*. Austin: University of Texas Press, 1992.

Blackburn, Simon. *The Oxford Dictionary of Philosophy*. New York: Oxford University Press, 1994.

Blackmore, Susan. *Consciousness: An Introduction*. New York: Oxford University Press, 2004.

Boeke, Kees. *Cosmic View: The Universe in 40 Jumps*. New York: John Day, 1957.

Bourget, David, and David J. Chalmers. "What Do Philosophers Believe?" *PhilPapers*, 2013. https://philpapers.org/archive/BOUWDP.pdf.

Brainard, F. Samuel. "Defining 'Mystical Experience.'" *Journal of the American Academy of Religion*, Summer 1996, 359–393.

———. *Reality and Mystical Experience*. University Park: Pennsylvania State University Press, 2000.

Brooks, Michael. *13 Things That Don't Make Sense*. New York: Doubleday, 2008.

Butchvarov, Panayot. "Metaphysical Realism." In *The Cambridge Dictionary of Philosophy*. 2nd ed. Edited by Robert Audi. New York: Cambridge University Press, 1999.

Campbell, Joseph. *The Masks of God: Oriental Mythology*. New York: Penguin, 1962.

Carmy, Shalom, and David Shatz. "The Bible as a Source for Philosophical Reflection." In *History of Jewish Philosophy*. Vol. 2 of Routledge History of World Philosophies, edited by Daniel H. Frank and Oliver Leaman. New York: Routledge, 2003.

Carroll, Sean. *The Higgs Boson and Beyond*. Course transcript. Chantilly, VA: The Teaching Company, 2015.

———. *Mysteries of Modern Physics: Time*. Course transcript. Chantilly, VA: The Teaching Company, 2012.

Carruthers, Peter. "Higher-Order Theories of Consciousness." In *The Stanford Encyclopedia of Philosophy* (Fall 2007 Edition), edited by Edward N. Zalta. Stanford University, 1997–. Article revised September 11, 2007. http://plato.stanford.edu/archives/fall2007/entries/ consciousness-higher/.

Chalmers, David J. "Consciousness." In *The Oxford Companion to the Mind*. 2nd ed. Edited by Richard L. Gregory, 207–208. New York: Oxford University Press, 2004.

———. "Facing Up to the Problem of Consciousness." *Journal of Consciousness Studies* 2, no. 3 (1995): 200–219. http://consc.net/papers/facing.html.

———. "How Do You Explain Consciousness?" TED talk, 2014. http://www.ted.com/talks/ david_chalmers_how_do_you_explain_consciousness?language=en.

Clayton, Philip, and Paul Davies, eds. *The Re-emergence of Emergence: The Emergentist Hypothesis from Science to Religion*. New York: Oxford University Press, 2006.

Clements, Ronald E. "Deuteronomy, The Book of." *The Oxford Companion to the Bible*. Edited by Bruce M. Metzger and Michael D. Coogan. New York: Oxford University Press, 1993.

Cohen, S. Marc. "Aristotle's Metaphysics." In *The Stanford Encyclopedia of Philosophy* (Spring 2009 Edition), edited by Edward N. Zalta. Stanford University, 1997–. Article revised June 9, 2008. http://plato.stanford.edu/archives/spr2009/entries/aristotle-metaphysics/.

Conio, Caterina. "Purāṇas." *The Encyclopedia of Religion*. Ed. in chief, Mircea Eliade. New York: Macmillan, 1987.

Conze, Edward, ed. and trans. *Buddhist Scriptures*. New York: Penguin, 1959.

Crane, Tim. "The Problem of Perception." In *The Stanford Encyclopedia of Philosophy* (Spring 2011 Edition), edited by Edward N. Zalta. Stanford University, 1997–. Article revised February 4, 2011. http://plato.stanford.edu/archives/spr2011/entries/perception-problem/.

Creath, Richard. "Logical Empiricism." In *The Stanford Encyclopedia of Philosophy* (Spring 2014 Edition), edited by Edward N. Zalta. Stanford University, 1997–. Article revised September 19, 2011. http://plato.stanford.edu/archives/spr2014/entries/logical-empiricism/.

Crowell, Steven. "Existentialism." In *The Stanford Encyclopedia of Philosophy* (Winter 2011 Edition), edited by Edward N. Zalta. Stanford University, 1997–. Article revised November 26, 2008. http://plato.stanford.edu/archives/win2008/entries/existentialism/.

De Muynck, Willem M. *Foundations of Quantum Mechanics: An Empiricist Approach*. New York: Kluwer Academic, 2002.

Dennett, Daniel C. "Consciousness." *The Oxford Companion to the Mind*. Edited by Richard L. Gregory, with the assistance of O. L. Zangwill. New York: Oxford University Press, 1987.

———. "Quining Qualia." In *The Nature of Consciousness: Philosophical Debates*. Edited by Ned Block, Owen Flanagan, and Güven Güzeldere, 619–642. Cambridge, MA: MIT Press, 1997.

Derrida, Jacques. *Margins of Philosophy*. Translated by Alan Bass. Chicago: University of Chicago Press, 1982.

Descartes, René. *Meditations on First Philosophy*. Translated by Donald A. Cress. In *Classics of Western Philosophy*. 2nd ed, edited by Steven M. Cahn. Indianapolis: Hackett, 1985.

Deutsch, Eliot. *Advaita-Vedānta: A Philosophical Reconstruction*. Honolulu: University of Hawaii Press, 1969.

De Wulf, M. "Roscelin." In *The Catholic Encyclopedia*. New York: Robert Appleton, 2014. http://www.newadvent.org/cathen/13189c.htm.

Dōgen. *Flowers of Emptiness: Selections from Dōgen's Shōbōgenzō*. Translated with introduction and notes by Hee-Jin Kim. New York: Edwin Mellen Press, 1985.

Downing, Lisa. "George Berkeley." In *The Stanford Encyclopedia of Philosophy* (Winter 2008 Edition), edited by Edward N. Zalta. Stanford University, 1997–. Article published September 10, 2004. http://plato.stanford.edu/archives/win2008/entries/berkeley/.

Dravid, Raja Ram. *The Problem of Universals in Indian Philosophy*. Delhi: Motilal Banarsidass, 1972.

Dretske, Fred. "Perception." In *The Cambridge Dictionary of Philosophy*. 2nd ed. Edited by Robert Audi. New York: Cambridge University Press, 1999.

Droege, Paula. "Higher-Order Theories of Consciousness." *The Internet Encyclopedia of Philosophy*. Edited by James Fieser and Bradley Dowden, 2005. http://www.iep.utm.edu/consc-hi/.

Durham, Ian T. "Oskar Klein." *MacTutor History of Mathematics Archive*. 2001. http://www-history.mcs.st-andrews.ac.uk/Biographies/Klein_Oskar.html.

Durham, John I. "Exodus, The Book of." *The Oxford Companion to the Bible.* Edited by Bruce M. Metzger and Michael D. Coogan. New York: Oxford University Press, 1993.

Eck, Diana L. *Encountering God: A Spiritual Journey from Bozeman to Banaras.* 2nd ed. Boston: Beacon Press, 2003.

Eddington, A. S. *The Nature of the Physical World.* New York: Macmillan, 1928.

Edelglass, William, and Jay Garfield, eds. *Buddhist Philosophy: Essential Readings.* New York: Oxford University Press, 2009.

Edgerton, Franklin. "Interpretation of the Bhagavad-gītā." In *The Bhagavad-gītā,* translated and interpreted by Franklin Edgerton, 103. Cambridge, MA: Harvard University Press, 1972.

Edie, James M. *Edmund Husserl's Phenomenology: A Critical Commentary.* Bloomington: Indiana University Press, 1987.

Einstein, Albert. "Geometry and Experience." *MacTutor History of Mathematics Archive.* 2007. Accessed April 28, 2016. http://www-groups.dcs.st-and.ac.uk/history/Extras/Einstein _geometry.html.

Emerton, J. A. "Names of God in the Hebrew Bible." *The Oxford Companion to the Bible.* Edited by Bruce M. Metzger and Michael D. Coogan. New York: Oxford University Press, 1993.

The Encyclopedia of Eastern Philosophy and Religion. Edited by Stephan Schuhmacher and Gert Woerner. Boston: Shambhala, 1989.

Enquist, Brian J., Bruce H. Tiffney, and Karl J. Niklas. "Metabolic Scaling and the Evolutionary Dynamics of Plant Size, Form, and Diversity: Toward a Synthesis of Ecology, Evolution, and Paleontology." *International Journal of Plant Sciences* 168, no. 5 (2007): 729–749. http://www.journals.uchicago.edu/cgi-bin/resolve?id=doi:10.1086/513479&erFrom= 7699279782161572027Guest.

Farah, Martha J. "Visual Perception and Visual Awareness after Brain Damage: A Tutorial Overview." In *The Nature of Consciousness: Philosophical Debates.* Edited by Ned Block, Owen Flanagan, and Güven Güzeldere, 203–236. Cambridge, MA: MIT Press, 1997.

Faye, Jan. "Copenhagen Interpretation of Quantum Mechanics." In *The Stanford Encyclopedia of Philosophy* (Fall 2014 Edition), edited by Edward N. Zalta. Stanford University, 1997–. Article revised July 24, 2014. http://plato.stanford.edu/archives/fall2014/entries/ qm-copenhagen/.

Feibleman, James K. "Real." *Dictionary of Philosophy.* Edited by Dagobert D. Runes. Totowa, NJ: Rowman and Allanheld, 1983.

Feynman, Richard. *The Character of Physical Law.* Cambridge, MA: MIT Press, 1993.

Filoramo, Giovanni. *A History of Gnosticism.* Cambridge, MA: Blackwell, 1990.

Fine, Cordelia. *A Mind of Its Own: How Your Brain Distorts and Deceives.* New York: W. W. Norton, 2006.

First Panchen Lama. *The Great Seal of Voidness.* Dharamsala, India: Library of Tibetan Works and Archives, 1975.

Fishbane, Michael. *Biblical Interpretation in Ancient Israel.* New York: Clarendon Press, 1985.

Fong, Benjamin Y. "On Critics and What's Real: Russell McCutcheon on Religious Experience." *Journal of the American Academy of Religion* 82, no. 4 (December 2014): 1127–1148.

Forbes, Graeme. "Reality." *The Cambridge Dictionary of Philosophy.* 2nd ed. Edited by Robert Audi. New York: Cambridge University Press, 1999.

Forman, Robert K. C., ed. *The Problem of Pure Consciousness*. New York: Oxford University Press, 1990.

Freud, Sigmund. *Civilization and Its Discontents*. New York: W. W. Norton, 1961.

Fulco, William J. "Baal." *The Encyclopedia of Religion*. Ed. in chief, Mircea Eliade. New York: Macmillan, 1987.

Fumerton, Robert A. "Logical Positivism." *The Cambridge Dictionary of Philosophy*. 2nd ed. Edited by Robert Audi. New York: Cambridge University Press, 1999.

Fung, Yu-lan. *A History of Chinese Philosophy, vol. 1*. Translated by Derk Bode. Princeton: Princeton University Press, 1952.

Garrett, Don. "Spinoza, Baruch." *The Cambridge Dictionary of Philosophy*. 2nd ed. Edited by Robert Audi. New York: Cambridge University Press, 1999.

Geertz, Clifford. *The Interpretation of Cultures: Selected Essays*. New York: Basic Books, 1973.

Gennaro, Rocco J. "Consciousness." *The Internet Encyclopedia of Philosophy*. Edited by James Fieser and Bradley Dowden. 2005. https://mitpress.mit.edu/books/consciousness -paradox.

Gillooly, James F., Andrew P. Allen, Geoffrey B. West, and James H. Brown. "The Rate of DNA Evolution: Effects of Body Size and Temperature on the Molecular Clock." *Proceedings of the National Academy of Sciences of the United States of America*. 102, no. 1 (2004): 140–145.

Gilmore, Cody. "Location and Mereology." In *The Stanford Encyclopedia of Philosophy* (Fall 2014 Edition), edited by Edward N. Zalta. Stanford University, 1997–. Article published June 7, 2013. https://plato.stanford.edu/archives/fall2014/entries/location-mereology/.

Goff, Philip. "Gerald Vision: *Re-emergence: Locating Conscious Properties in a Material World*." Book review in *Notre Dame Philosophical Reviews* 4, no. 30 (2012). https://ndpr.nd.edu/ news/30555-re-emergence-locating-conscious-properties-in-a-material-world/.

Gold, Daniel. "Consecration." *The Encyclopedia of Religion*. Ed. in chief, Mircea Eliade. New York: Macmillan, 1987.

Goldman, Steven L. *Science in the Twentieth Century: A Social-Intellectual Survey*. Course transcript (3 volumes). Chantilly, VA: The Teaching Company, 2004.

———. *Science Wars: What Scientists Know and How They Know It*. Course transcript (2 volumes). Chantilly, VA: The Teaching Company, 2006.

Goodman, Nelson. *Fact, Fiction, and Forecast*. 4th ed. Cambridge, MA: Harvard University Press, 1983.

Graham, A. C. *Disputers of the Tao*. La Salle, IL: Open Court, 1989.

Greene, Brian. *The Elegant Universe*. New York: Vintage Books, 1999.

Gregory, Richard. "Perception as Hypotheses." *The Oxford Companion to the Mind*. Edited by Richard L. Gregory, with the assistance of O. L. Zangwill. New York: Oxford University Press, 1987.

———. "Perception as Unconscious Influences." *The Oxford Companion to the Mind*. 2nd ed. Edited by Richard L. Gregory. New York: Oxford University Press, 2004.

Gribbin, John. *In Search of Schrödinger's Cat: Quantum Physics and Reality*. New York: Bantam Books, 1984.

Grier, Michelle. "Kant's Critique of Metaphysics." In *The Stanford Encyclopedia of Philosophy* (Summer 2012 Edition), edited by Edward N. Zalta. Stanford University, 1997–.

Article revised April 10, 2012. http://plato.stanford.edu/archives/sum2012/entries/kant-metaphysics/.

Grim, Patrick. *Philosophy of Mind: Brains, Consciousness, and Thinking Machines.* Chantilly, VA: The Teaching Company, 2008.

Guthrie, W. K. C. *A History of Greek Philosophy.* New York: Cambridge University Press, 1969.

Guttenplan, Samuel, ed. *A Companion to the Philosophy of Mind.* Blackwell Companions to Philosophy. Cambridge, MA: Blackwell, 1995.

Güzeldere, Güven. "Introduction: The Many Faces of Consciousness: A Field Guide." In *The Nature of Consciousness: Philosophical Debates.* Edited by Ned Block, Owen Flanagan, and Güven Güzeldere, 1–68. Cambridge, MA: MIT Press, 1997.

Halpern, Baruch. "Monotheism." *The Oxford Companion to the Bible.* Edited by Bruce M. Metzger and Michael D. Coogan. New York: Oxford University Press, 1993.

Hanh, Thich Nhat. *Touching Peace: Practicing the Art of Mindful Living.* Berkeley, CA: Parallax Press, 1992.

Haq, Syed Nomanul. "The Indian and Persian Background." In *History of Islamic Philosophy,* edited by Seyyed Hossein Nasr and Oliver Leaman, 52–70. New York: Routledge, 2001.

Harrison, David M. "Basal Metabolic Rate." *Upscale.* Accessed April 28, 2016. http://www.upscale.utoronto.ca/PVB/Harrison/BasalMetabolism/BasalMetabolism.html2006.

Hart, William. *The Art of Living: Vipassana Meditation as Taught by S. N. Goenka.* New York: HarperCollins, 1987.

Hartshorne, Charles, and William L. Reese. *Philosophers Speak of God.* 2nd ed. Amherst, NY: Humanity Books, 2000.

Harvey, Peter. *An Introduction to Buddhism: Teachings, History, and Practices.* Cambridge: Cambridge University Press, 1990.

Healy, K., L. McNally, G. D. Ruxton, N. Cooper, and A. L. Jackson. "Metabolic Rate and Body Size Are Linked with Perception of Temporal Information." *Animal Behaviour* 86, no. 4 (2013): 685–696. http://www.ncbi.nlm.nih.gov/pmc/articles/PMC3791410/.

Heidegger, Martin. *Being and Time.* Translated by John Macquarrie and Edward Robinson. New York: Harper and Row, 1962.

Herman, A. L. *An Introduction to Buddhist Thought: A Philosophical History of Indian Buddhism.* Lanham: University Press of America, 1983.

Hick, John. *Philosophy of Religion.* 3rd ed. Englewood Cliffs, NJ: Prentice-Hall, 1983.

Hiltebeitel, Alf. "Hinduism." *The Encyclopedia of Religion.* Ed. in chief, Mircea Eliade. New York: Macmillan, 1987.

Hofstadter, Douglas R. *Gödel, Escher, Bach: An Eternal Golden Braid.* New York: Vintage Books, 1980.

Holt, Jim. "Mind of a Rock: Is Everything Conscious?" *New York Times Magazine,* November 18, 2007, section 6, 19–20.

Holton, Gerald, and Stephen G. Brush. *Introduction to Concepts and Theories in Physical Science.* 2nd ed. Princeton: Princeton University Press, 1985.

Horton, Robert. "African Traditional Thought and Western Science." In *Rationality,* edited by Bryan R. Wilson, 131–171. Oxford: Basil Blackwell, 1970.

Horwich, Paul. "Truth." *The Cambridge Dictionary of Philosophy.* 2nd ed. Edited by Robert Audi. New York: Cambridge University Press, 1999.

Hossenfelder, Sabine. "10 Quantum Truths about Our Universe." *Forbes*, March 15, 2016. http://www.forbes.com/sites/startswithabang/2016/03/15/10-quantum-truths-about-our -universe/#410793d74642.

Hume, David. *An Enquiry concerning Human Understanding*. In *Classics of Western Philosophy: Second Edition*. Edited by Steven M. Cahn. Indianapolis: Hackett, 1985.

Husserl, Edmund. *Cartesian Mediations: An Introduction to Phenomenology*. Translated by Dorion Cairns. Boston: Martinus Nijhoff, 1960.

———. *Ideas: General Introduction to Pure Phenomenology*. Translated by W. R. Boyce Gibson. New York: Collier, 1962.

Hyslop, Alec. "Other Minds." In *The Stanford Encyclopedia of Philosophy* (Fall 2008 Edition), edited by Edward N. Zalta. Stanford University, 1997–. Article published October 6, 2005. http://plato.stanford.edu/archives/fall2008/entries/other-minds/.

Ibn-al'Arabī. *The Bezels of Wisdom*. Translated by R. W. J. Austin. The Classics of Western Spirituality. New York: Paulist Press, 1980.

Idel, Moshe. *Kabbalah: New Perspectives*. New Haven: Yale University Press, 1988.

Idel, Moshe, and Bernard McGinn, eds. *Mystical Union and Monotheistic Faith: An Ecumenical Dialogue*. New York: Macmillan, 1989.

Jacobsen, Thorkild. "Mesopotamian Religions: An Overview." *The Encyclopedia of Religion*. Ed. in chief, Mircea Eliade. New York: Macmillan, 1987.

James, William. *The Varieties of Religious Experience*. New York: Collier, 1961.

Jaspers, Karl. *Origin and Goal of History*. New York: Routledge Revivals, 2010.

Jayatilleke, K. N. *Early Buddhist Theory of Knowledge*. Delhi: Motilal Banarsidass, 1980.

Jech, Thomas. "Set Theory." In *The Stanford Encyclopedia of Philosophy* (Summer 2011 Edition), edited by Edward N. Zalta. Stanford University, 1997–. Article published July 11, 2002. http://plato.stanford.edu/archives/sum2011/entries/set-theory/.

Johnson, David Alan. "Grue Paradox." *The Cambridge Dictionary of Philosophy*. 2nd ed. Edited by Robert Audi. New York: Cambridge University Press, 1999.

Johnson, George. "Can a String of Numbers Tie the Universe Together?" *New York Times*, February 5, 1995. Ideas and Trends: Science, E3.

Jordan, Michael. *Encyclopedia of Gods*. New York: Facts on File, 1993.

Jung, C. G. *Two Essays on Analytical Psychology*. 2nd. ed. Translated by R. F. C. Hull. Princeton: Princeton University Press, 1966.

Kafatos, Menas, and Robert Nadeau. *The Conscious Universe: Part and Whole in Modern Physical Theory*. New York: Springer-Verlag, 1990.

Kahane, Howard. *Logic and Philosophy: A Modern Introduction*. 5th ed. Belmont, CA: Wadsworth, 1986.

Kalupahana, David J. "Pratītya-Samutpāda." *The Encyclopedia of Religion*. Ed. in chief, Mircea Eliade. New York: Macmillan, 1987.

Kane, Gordon. "The Dawn of Physics beyond the Standard Model." *Scientific American* 288 (June 2003): 68–75.

Kant, Immanuel. *Critique of Pure Reason*. Translated by Norman Kemp Smith. New York: St. Martin's Press, 1965.

Karlen, Peter H. "Humor and Enlightenment, Part I: The Theory." *Contemporary Aesthetics* 14 (March 29, 2016).

Keil, Geert. "Making Something Happen: Where Causation and Agency Meet." In *Agency and Causation in the Human Sciences*, edited by Josef Quitterer et al. Paderborn (mentis) 2007, 19–35. Reprinted in *Nature and Rational Agency*, edited by Siri Granum Carson and Kjartan Mikalsen Koch, 9–28. Frankfurt am Main: Berlin Bern Bruxelles, 2009. https://www.philosophie.hu-berlin.de/de/lehrbereiche/anthro/mitarbeiter/keil/pdfs/c32volltext _reprint.pdf.

Kelly, Thomas. "Evidence." In *The Stanford Encyclopedia of Philosophy* (Fall 2012 Edition), edited by Edward N. Zalta. Stanford University, 1997–. Article revised July 28, 2014. http://plato.stanford.edu/archives/fall2014/entries/evidence/.

Kim, Jaegwon. "Causation." *The Cambridge Dictionary of Philosophy*. 2nd ed. Edited by Robert Audi. New York: Cambridge University Press, 1999.

———. "Explanation." *The Cambridge Dictionary of Philosophy*. 2nd ed. Edited by Robert Audi. New York: Cambridge University Press, 1999.

King, Richard. *Indian Philosophy: An Introduction to Hindu and Buddhist Thought*. Washington, DC: Georgetown University Press, 1999.

Klarreich, Erica. "Life on the Scales: Simple Mathematical Relationships Underpin Much of Biology and Ecology." *Science News* 167 (February 12, 2005): 106–108.

Klein, Anne C. *Knowledge and Liberation: Tibetan Buddhist Epistemology in Support of Transformative Religious Experience*. Ithaca, NY: Snow Lion, 1986.

Kohák, Erazim. *Idea and Experience: Edmund Husserl's Project of Phenomenology in Ideas I*. Chicago: University of Chicago Press, 1978.

Körner, Stephen. *Metaphysics: Its Structure and Function*. New York: Cambridge University Press, 1984.

Krauss, Lawrence M. "The Trouble with Theories of Everything." *Nautilus*, October 1. 2015. Accessed April 28, 2016. http://nautil.us/issue/29/scaling/the-trouble-with-theories-of -everything.

Laumakis, Stephen J. *An Introduction to Buddhist Philosophy*. New York: Cambridge University Press, 2008.

Libet, B. "Unconscious Cerebral Initiative and the Role of Conscious Will in Voluntary Action." *Behavioral and Brain Sciences* 8 (1985): 529–566.

Lipner, Julius. *Hindus: Their Religious Beliefs and Practices*. New York: Routledge, 1994.

Lloyd, Elisabeth. "Units and Levels of Selection." In *The Stanford Encyclopedia of Philosophy* (Fall 2007 Edition), edited by Edward N. Zalta. Stanford University, 1997–. Article revised April 14, 2017. http://plato.stanford.edu/entries/selection-units/.

Locke, John. *An Essay concerning Human Understanding*. Edited by Roger Woolhouse. New York: Penguin, 2004.

Look, Brandon C. "Gottfried Wilhelm Leibniz." In *The Stanford Encyclopedia of Philosophy* (Spring 2014 Edition), edited by Edward N. Zalta. Stanford University, 1997–. Article revised July 24, 2013. http://plato.stanford.edu/archives/spr2014/entries/leibniz/.

Louth, Andrew. *The Origins of the Christian Mystical Tradition: From Plato to Denys*. Oxford: Clarendon, 1981.

Ludwig, Theodore M. "Gods and Goddesses." *The Encyclopedia of Religion*. Ed. in chief, Mircea Eliade. New York: Macmillan, 1987.

———. *The Sacred Paths of the East*. Upper Saddle River, NJ: Prentice Hall, 1993.

Lusthaus, Dan. *Buddhist Phenomenology: A Philosophical Investigation of Yogācāra Buddhism and the Ch'eng Wei-shih lun.* New York: RoutledgeCurzon, 2002.

Maddy, Penelope. "Set Theory." *The Cambridge Dictionary of Philosophy.* 2nd ed. Edited by Robert Audi. New York: Cambridge University Press, 1999.

Maimonides, Moses. *The Guide for the Perplexed.* Translated by M. Friedländer. New York: Dover, 1956.

Matilal, Bimal Krishna. *Logic, Language and Reality: Indian Philosophy and Contemporary Issues.* Delhi: Motilal Banarsidass, 1990.

———. *Perception: An Essay on Classical Indian Theories of Knowledge.* New York: Oxford University Press, 1986.

May, Ernest D. *The Harvard Dictionary of Music, 4th Edition.* Edited by Don Michael Randel. Cambridge, MA: Belknap Press of Harvard University Press, 2003.

McGinn, Bernard. *The Foundations of Mysticism.* Vol. 1 of *The Presence of God: A History of Western Christian Mysticism.* New York: Crossroad, 1992.

———. "Love, Knowledge, and *Unio Mystica* in the Western Christian Tradition." In *Mystical Union and Monotheistic Faith: An Ecumenical Dialogue,* edited by Moshe Idel and Bernard McGinn, 59–86. New York: Macmillan, 1989.

Milius, Susan. "Predator-Prey Relationship Quantified: Power Law for Animal Abundances Matches '3/4 Rule' in Physiology." *Science News* 188, no. 7 (October 3, 2015).

Mill, John Stuart. *A System of Logic, Ratiocinative and Inductive.* 8th ed. New York: Harper & Brothers, 1982.

Morreall, John. "Philosophy of Humor." In *The Stanford Encyclopedia of Philosophy* (Spring 2013 Edition), edited by Edward N. Zalta. Stanford University, 1997–. Article published November 20, 2012. http://plato.stanford.edu/archives/spr2013/entries/humor/.

Moskowitz, Clara. "Tangled Up in Spacetime." *Scientific American* 316 (January 2017): 32–37.

Mother Teresa. *A Simple Path.* Compiled by Lucinda Vardey. New York: Ballantine, 1995.

Nāgārjuna. *Mūlamadhyamakakārikā.* Translated by with annotation by David J. Kalupahana under the title *Nāgārjuna: The Philosophy of the Middle Way.* Albany: SUNY Press, 1986.

Nagel, Thomas. *Mind and Cosmos: Why the Materialist Neo-Darwinian Conception of Nature Is Almost Certainly False.* New York: Oxford University Press, 2012.

———. *The View from Nowhere.* New York: Oxford University Press, 1986.

———. "What Is It like to Be a Bat?" *Philosophical Review* 4 (1974): 435–450.

Newberg, Andrew. "Are We 'Hard-Wired' for God?" *Andrew Newberg* (blog). September 26, 2013. http://www.andrewnewberg.com/research-blog/2013/9/26/are-we-hard-wired-for-god.

Niehoff, Debra. *The Language of Life: How Cells Communicate in Health and Disease.* Washington, DC: Joseph Henry Press, 2005.

Nikhilananda, Swami, trans. *The Upanishads.* New York: Bell, 1963.

Ninan, M. M. "Puranas and Their Dates." *Oration.com.* 2012. http://www.oration.com/~mm9n/articles/dev/07Puranas.htm.

Nottale, Laurent. *Fractal Space-Time and Microphysics: Toward a Theory of Scale Relativity.* Hackensack, NJ: World Scientific, 1993.

O'Brien, Elmer, trans. *The Essential Plotinus.* Indianapolis: Hackett, 1964.

O'Connor, J. J., and E. F. Robertson. "A History of Set Theory." *Philosophical Explorations.* 1996. http://braungardt.trialectics.com/mathematics/mathematicians/cantor/set-theory/.

———. "Theodor Franz Eduard Kaluza." *MacTutor History of Mathematics Archive*. 2000. Accessed April 28, 2016. http://www-groups.dcs.st-and.ac.uk/history/Biographies/ Kaluza.html.

O'Connor, Timothy, and Hong Yu Wong. "Emergent Properties." In *The Stanford Encyclopedia of Philosophy* (Summer 2015 Edition), edited by Edward N. Zalta. Stanford University, 1997–. Article revised June 3, 2015. http://plato.stanford.edu/archives/sum2015/entries/ properties-emergent/.

Passmore, John. "Logical Positivism." *The Encyclopedia of Philosophy*. Ed. in chief, Paul Edwards. New York: Collier Macmillan, 1967.

Peirce, Charles Sanders. *Philosophical Writings of Peirce*. Edited by Justus Buchler. New York: Dover, 1955.

Plato. *Phaedo*. In *Five Dialogues*, translated by G. M. A. Grube. Indianapolis: Hackett, 1981.

———. *Plato's Republic*. Translated by G. M. A. Grube. Indianapolis: Hackett, 1974.

———. *Timaeus*. In *The Dialogues of Plato*, translated by B. Jowett. New York: Random House, 1937.

Prebish, Charles S. "Doctrines of the Early Buddhists." In *Buddhism: A Modern Perspective*, edited by Charles S. Prebish, 29–35. University Park: Pennsylvania State University Press, 1975.

Price, Momoko. "The Controversy of Group Selection Theory." *Science Creative Quarterly*, September 2007–April 2008, 1. http://www.scq.ubc.ca/the-controversy-of-group-selection -theory/.

Quartz, Steven R., and Terrence J. Sejnowski. *Liars, Lovers, and Heroes: What the New Brain Science Reveals about How We Become Who We Are*. New York: HarperCollins, 2002.

Quispel, Gilles. "Gnosticism: Gnosticism from Its Origin to the Middle Ages." *The Encyclopedia of Religion*. Ed. in chief, Mircea Eliade. New York: Macmillan, 1987.

Qur'an. Translated by Majid Fakhry. *An Interpretation of the Qur'an: English Translation of the Meanings; A Bilingual Edition*. New York: New York University Press, 2004.

Raatikainen, Panu. "Gödel's Incompleteness Theorems." In *The Stanford Encyclopedia of Philosophy* (Spring 2014 Edition), edited by Edward N. Zalta. Stanford University, 1997–. Article published November 11, 2013. http://plato.stanford.edu/archives/spr2014/entries/ goedel-incompleteness/.

Radhakrishnan, Sarvepalli, and Charles A. Moore, ed. *A Sourcebook in Indian Philosophy*. Princeton: Princeton University Press, 1957.

Rahula, Walpola. *What the Buddha Taught*. Revised Edition. New York: Grove Press, 1974.

Raju, P. T. *Structural Depths of Indian Thought*. Albany: SUNY Press, 1985.

Rea, Michael C., ed. *Material Constitution: A Reader*. New York: Rowman and Littlefield, 1997.

Reber, Arthur S. *The Penguin Dictionary of Psychology*. New York: Viking Penguin, 1985.

Reps, Paul, ed. *Zen Flesh, Zen Bones: A Collection of Zen and Pre-Zen Writings*. New York: Doubleday, 1987.

Rescher, Nicholas. "Idealism." In *The Oxford Companion to the Mind*. 2nd ed. Edited by Robert Audi. New York: Cambridge University Press, 1999.

Rey, Georges. "A Question about Consciousness." In *The Nature of Consciousness: Philosophical Debates*, edited by Ned Block, Owen Flanagan, and Güven Güzeldere, 461–482. Cambridge, MA: MIT Press, 1997.

Ricoeur, Paul. *Husserl: An Analysis of His Phenomenology.* Translated by Edward B. Ballard and Lester E. Embree. Evanston: Northwestern University Press, 1967.

Rizzolatti, Giacomo, and Laila Craighero. "The Mirror-Neuron System." *Annual Review of Neuroscience* 27 (2004): 169–192.

Robinson, Daniel N. *Consciousness and Its Implications.* Course transcript. Chantilly, VA: The Teaching Company, 2007.

Robinson, Richard, and Willard L. Johnson. *The Buddhist Religion: A Historical Introduction.* 4th ed. Belmont, CA: Wadsworth, 1997.

Rorty, Richard. *Philosophy and the Mirror of Nature.* Princeton: Princeton University Press, 1979.

———. "Relations, Internal and External." *The Encyclopedia of Philosophy.* New York: Macmillan, 1967.

Rosenberg, Alex. *The Philosophy of Science: A Contemporary Introduction.* New York: Routledge, 2000.

Rundle, Bede. "Logic: History of." *The Encyclopedia of Philosophy.* New York: Macmillan, 1967.

Sanders, Laura. "Mirror System Gets an Assist: Brain Module Gets Help when Actions Can't Be Mimicked." *Science News* 180, no. 4 (August 13, 2011): 2–32.

Śaṅkara. *The Vedānta Sūtras with the Commentary by Śakarākārya.* Translated by George Thibaut. Sacred Books of the East, 34 and 38. Oxford: Clarendon, 1890, 1896. Excerpted in *A Sourcebook in Indian Philosophy*, edited by Sarvepalli Radhakrishnan and Charles A. Moore. Princeton: Princeton University Press, 1957.

Sapire, David. "Disposition." In *The Cambridge Dictionary of Philosophy.* 2nd ed. Edited by Robert Audi. New York: Cambridge University Press, 1999.

Sarup, Madan. *An Introductory Guide to Post-Structuralism and Postmodernism.* 2nd. ed. Athens: University of Georgia Press, 1993.

Sawyer, John F. A. "Isaiah, The Book of." *The Oxford Companion to the Bible.* Edited by Bruce M. Metzger and Michael D. Coogan. New York: Oxford University Press, 1993.

Schimmel, Annemarie. *Mystical Dimensions of Islam.* Chapel Hill: University of North Carolina Press, 1975.

Schoedinger, Andrew B., ed. *The Problem of Universals.* Atlantic Highlands, NJ: Humanities Press, 1992.

Seager, William, and Sean Allen-Hermanson. "Panpsychism." In *The Stanford Encyclopedia of Philosophy* (Spring 2012 Edition), edited by Edward N. Zalta. Stanford University, 1997–. Article revised August 23, 2010. http://plato.stanford.edu/archives/spr2012/entries/panpsychism/.

Sedley, David N. "Stoicism." In *The Cambridge Dictionary of Philosophy.* 2nd ed. Edited by Robert Audi. New York: Cambridge University Press, 1999.

Seligman, Martin E. P. *Authentic Happiness: Using the New Positive Psychology to Realize Your Potential for Lasting Fulfillment.* New York: Free Press, 2002.

Sells, Michael A. *Mystical Languages of Unsaying.* Chicago: University of Chicago Press, 1994.

Shallice, Tim. "Modularity and Consciousness." In *The Nature of Consciousness: Philosophical Debates.* Edited by Ned Block, Owen Flanagan, and Güven Güzeldere, 255–276. Cambridge, MA: MIT Press, 1997.

Shaner, David Edward. *The Bodymind Experience in Japanese Buddhism: A Phenomenological Study of Kūkai and Dōgen.* Albany: SUNY Press, 1985.

Shoemaker, Sydney. "Physicalism." In *The Cambridge Dictionary of Philosophy*. 2nd ed. Edited by Robert Audi. New York: Cambridge University Press, 1999.

Sidharth, B. G. *The Thermodynamic Universe: Exploring the Limits of Physics*. Hackensack, NJ: World Scientific, 2008.

Siewert, Charles. "Consciousness and Intentionality." In *The Stanford Encyclopedia of Philosophy* (Fall 2008 Edition), edited by Edward N. Zalta. Stanford University, 1997–. Article revised December 23, 2006. http://plato.stanford.edu/archives/fall2008/entries/consciousness-intentionality/.

Skinner, B. F. "'Superstition' in the Pigeon." *Journal of Experimental Psychology* 38 (1947).

Sklar, Lawrence. *Philosophy of Physics*. Boulder: Westview, 1992.

Skorupski, Tadeusz. "Dharma: Buddhist Dharma and Dharmas." *The Encyclopedia of Religion*. Ed. in chief, Mircea Eliade. New York: Macmillan, 1987.

Slobodchikoff, Con. *Chasing Doctor Dolittle: Learning the Language of Animals*. New York: St. Martin's Press, 2012.

Smith, David Woodruff. "Phenomenology." In *The Stanford Encyclopedia of Philosophy* (Summer 2009 Edition), edited by Edward N. Zalta. Stanford University, 1997–. Article revised July 28, 2008. http://plato.stanford.edu/archives/sum2009/entries/phenomenology/.

Smith, Huston. *The World's Religions*. New York: HarperCollins, 1991.

Smolin, Lee. *Three Roads to Quantum Gravity*. New York: Basic Books, 2001.

Smuts, Aaron. "Humor." *Internet Encyclopedia of Philosophy*. 2009. http://www.iep.utm.edu/humor/.

Sperling, S. David. "God: God in the Hebrew Scriptures." *The Encyclopedia of Religion*. Ed. in chief, Mircea Eliade. New York: Macmillan, 1987.

Staal, Frits. *Universals: Studies in Indian Logic and Linguistics*. Chicago: University of Chicago Press, 1998.

Stalker, Douglas, ed. *Grue! The New Riddle of Induction*. Chicago: Open Court, 1994.

Swoyer, Chris, and Francesco Orilia. "Properties." In *The Stanford Encyclopedia of Philosophy* (Fall 2011 Edition), edited by Edward N. Zalta. Stanford University, 1997–. Article revised September 12, 2011.http://plato.stanford.edu/archives/fall2011/entries/properties/.

Ulin, Robert Charles. *Understanding Cultures: Perspectives in Anthropology and Social Theory*. Malden, MA: Wiley-Blackwell, 2001.

Unno, Taitetsu. "Eightfold Path." *The Encyclopedia of Religion*. Ed. in chief, Mircea Eliade. New York: Macmillan, 1987.

Van Gulick, Robert. "Consciousness." In *The Stanford Encyclopedia of Philosophy* (Spring 2014 Edition), edited by Edward N. Zalta. Stanford University, 1997–. Article revised January 14, 2014. http://plato.stanford.edu/archives/spr2014/entries/consciousness/.

Van Voorst, Robert. *Anthology of Asian Scriptures*. Belmont, CA: Wadsworth, 2001.

Varzi, Achille. "Mereology." In *The Stanford Encyclopedia of Philosophy* (Winter 2016 Edition), edited by Edward N. Zalta. Stanford University, 1997–. Article revised February 13, 2016. http://plato.stanford.edu/archives/win2016/entries/mereology/.

Vickers, John. "The Problem of Induction." In *The Stanford Encyclopedia of Philosophy* (Fall 2011 Edition), edited by Edward N. Zalta. Stanford University, 1997–. Article revised June 21, 2010. http://plato.stanford.edu/archives/fall2011/entries/induction-problem/.

Vision, Gerald. *Re-emergence: Locating Conscious Properties in a Material World.* Cambridge, MA: MIT Press, 2011.

Vivekananda, Swami. *Jnāna-Yoga.* Revised Edition. New York: Ramakrishna-Vivekananda Center, 1955.

———. *Karma-Yoga and Bhakti-Yoga.* Revised Edition. New York: Ramakrishna-Vivekananda Center, 1955.

———. *Rāja-Yoga.* Revised Edition. New York: Ramakrishna-Vivekananda Center, 1955.

Vogt, Katja. "Ancient Skepticism." In *The Stanford Encyclopedia of Philosophy* (Fall 2015 Edition), edited by Edward N. Zalta. Stanford University, 1997–. Article revised May 31, 2014. http://plato.stanford.edu/archives/fall2015/entries/skepticism-ancient/.

Von Baeyer, Hans Christian. "World on a String." *The Sciences*, September/October 1999, 10–13.

Waldron, William S. *The Buddhist Unconscious: The Alaya-vijñāna in the Context of Indian Buddhist Thought.* Routledge Critical Studies in Buddhism. New York: RoutledgeCurzon, 2003.

Wallis, Claudia. "The New Science of Happiness." *Time*, January 17, 2005, A3–A9.

Warder, A. K. *Indian Buddhism.* Delhi: Motilal Banarsidass, 1980.

Wehner, Stephanie. "Two Quantum Mysteries Merged into One." *Centre for Quantum Technologies.* 2014. http://www.quantumlah.org/highlight/141220_wave_particle.php.

Weinberg, Steven. *The First Three Minutes: A Modern View of the Origin of the Universe.* Updated Edition. New York: Basic Books, 1993.

Weinfeld, Moshe. "Israelite Religion." *The Encyclopedia of Religion.* Ed. in chief, Mircea Eliade. New York: Macmillan, 1987.

West, Geoffrey B., and James H. Brown. "Life's Universal Scaling Laws." *Physics Today*, September 2004. http://www.physycom.unibo.it/bazzani_web_dir/sistemi_complessi/scaling.pdf.

Wetzel, Linda. "Types and Tokens." In *The Stanford Encyclopedia of Philosophy* (Winter 2008 Edition), edited by Edward N. Zalta. Stanford University, 1997–. Article published April 28, 2006. http://plato.stanford.edu/archives/win2008/entries/types-tokens/.

White, Sidnie Ann. "Human Person." *The Oxford Companion to the Bible.* Edited by Bruce M. Metzger and Michael D. Coogan. New York: Oxford University Press, 1993.

Whitehead, Alfred North. *Process and Reality.* Corrected edition. Edited by David Ray Griffin and Donald W. Sherburne. New York: Free Press, 1978.

Whitfield, John. "All Creatures Great and Small." *Nature* 413, no. 6854 (September 27, 2001): 342–344. http://www.fractal.org/Life-Science-Technology/Publications/All-creatures.htm.

Wigner, Eugene P. "The Unreasonable Effectiveness of Mathematics in the Natural Sciences." Richard Courant Lecture in Mathematical Sciences delivered at New York University, May 11, 1959. *Communications on Pure and Applied Mathematics* 13, no. 1 (1960).

Wittgenstein, Ludwig. *Philosophical Investigations.* The English text of the 3rd ed. Translated by G. E. M. Anscombe. New York: Macmillan, 1958.

Wood, Ledger. "Publicity, Epistemic." *Dictionary of Philosophy.* Edited by Dagobert D. Runes. Totowa, NJ: Rowman and Allanheld, 1983.

Woozley, A. D. "Universals." *The Encyclopedia of Philosophy.* New York: Macmillan, 1967.

Wright, Robert. *The Evolution of God.* New York: Black Bay Books, 2009.

Index

Notes are indicated by n.

awareness (*continued*)

as collective act, 138–40, 226n12

conjunctive role of, 127, 130–31, 132, 135–36, 138, 149, 150, 225n7

and consciousness, 65–68, 132–35

definition of, 137–39, 146, 149, 156

disjunctive role of, 126, 127, 130–31, 132, 135–36, 138, 149, 150, 225n7, 226n13

as distinguishing human beings from matter, 147, 151–53

dual roles of, 126–29, 172–74, 225n7, 225–26n9

and emergence, 153–54

and existence, 139–40

first-person, 126–28, 139, 169

material, 162–63, 165

and presence, 127, 129–32, 137–39, 149, 177, 219n8, 226n10

and publicities, 135, 137–39, 153–54, 234n4, 234n4

and reality, 137

and the self, 67, 148–49

and set theory, 156, 178

and virtue, 177

See also conscious awareness; consciousness; dhammas; presence; unconscious awareness

Axial Age, 30, 208n8

Baars, Bernard J., 214n17

Baeyer, Hans Christian van, 158–59

Barnes, Jonathan, 208n6

Barrow, John D., 207n16

Basil of Caesarea, 224n25

Berkeley, George, 40, 213n11, 214n12, 216n10

Bhagavad-gītā, 47

bhakti yoga, 51–52, 55

biology, and scale, 184–85. *See also* cells

Blackburn, Simon, 206n12, 207n13, 219n14

Blackmore, Susan, 225n2

Blavatsky, H. P., 210n2

bodhisattva, 83, 175

Bohr, Niels, 19–20, 207n16

Boltzmann, Ludwig, 186

Bonaventure, Saint, 59

Brahma, 44

Brahman, 30, 38, 44–48, 50, 53–54, 175, 210n6, 211n9

archetypal attributes of, 47–48

as archetype of the aware self, 45–48, 54–55, 57, 62, 65, 167, 211n15

concept of, 43, 45–47, 52–53, 211n10

five sheaths of, 52–53, 173

and Neoplatonism, 113

as ultimate reality, 45, 56, 212n18

See also God

brain

and meditation, 235n21

versus mind, 17, 134–35

scientific investigation of, 134–35

Brentano, Franz, 225n6

Buddha, the (Siddhartha Gautama), 30, 69, 85, 215n4

traditional story of, 70–71

Buddhism, 4–5, 30, 173, 233n2

and awareness, 67, 215n8

Buddha's death, after the, 82–84

causality as aspect of, 73, 74–75, 76–78, 86, 97, 216n13

emptiness as aspect of, 83–84

as first-person account of reality, 72–73, 76, 84, 86, 88, 169, 216n9

Four Noble Truths of, 71–72, 75, 85, 213n7

free will in, 223n14

impermanence as aspect of, 72–73, 216n13

interdependent arising in, 73, 74–75, 76, 85, 215n7, 217n17

and material properties, 76–78

and meditation as spiritual practice, 175

and phenomenology, 213n7, 215n5, 218n26

philosophical difficulties inherent in, 78, 86, 216–17n13

presence as aspect of, 80, 82, 168

sects of, 83–84, 216–17n13

and the self, 72–73, 124, 126, 226n13, 233n2

spiritual practices in, 72, 79–80, 84, 167–68, 175, 215n8, 217nn17–18
and suffering (dukkha), 71, 75, 85, 173, 179
Three Jewels of, 215n4
Buddhist Wheel of Life, 74–75

Campbell, Joseph, 210n6
Carnap, Rudolf, 6, 218–19n2
Carroll, Sean, 237–38n9
causality, 219–20n14, 220n15
and agency, 96–97
as aspect of reality, 65, 171
atoms as example of, 35, 96, 209n18
and Buddhism, 73, 74–75, 76–78
cells
and awareness, 153–54
as collectivities of reference, 144–46
Chalmers, David, 134, 227n19
Chandrakīrti, 84
change
in Buddhism, 30
Heraclitus's view of, 30–31
Parmenides's view of, 31
as ultimate reality, 5, 7
and universals, 28, 171, 208n2
chit (consciousness), 47–48
Chomsky, Noam, 225
Christianity
and concept of God, 113, 176
mysticism in, 113, 224n25
and union with God, 224n28
See also Abrahamic religions; Western monotheism
classes, as distinguished from CORs, 140, 150
class logic, 36, 37. See also set theory
Clayton, Philip, 230n4
cognition, 31
Cohen, S. Mark, 209n12
collectivities of reference (CORs), 140–47
cells as, 144–46
as distinguished from classes, 140, 141, 143, 150
and emergence, 152–53

human beings as, 140, 141–42, 143–44, 228n7
and mathematics, 155–56
matter as, 147, 228n7, 232n29
and the mind-matter distinction, 147–48
and personal identity, 148–49
primary and secondary, 228n5, 232n29
and reality, 164, 165, 238n5
and scale, 185–86
and time and space, 183–84, 233n30
and truth, 188–90
common sense realism (theory of perception), 206n10
communities of discourse (CODs), 146–47, 150
compassion, 177, 217n17
conceptualism, 40
conscious awareness, 65–68, 76, 126, 132, 136, 214nn16–17, 228n2
consciousness, 13
and awareness, 65–68, 132–35, 142–43, 213n11
Berkeley's view of, 213n11
Buddhist concept of, 75–76, 84
higher-order theories of, 66–67, 68, 214n16
Hindu concept of, 47, 55
Husserl's view of, 213n11
intransitive versus transitive, 133
problem of, 134–35, 136, 151, 205n4, 227n19, 227n22
pure, 133
in Yogācāra Buddhism, 84
constructivists, 205n3
cosmos
as governed by universal properties, 30–31, 94–95
Greek concept of, 94–95
counterfactual conditionals, 93
Cratylus, 57
creation stories
in Hinduism, 45–26, 210n6
in Judeo-Christian tradition, 45

cultural context, as influence on our understanding of the world, 125–26

Dao, 30
Daoism, 5, 177
Davies, Paul, 230n4
death, understanding of, 236n30
de Compiègne, Roscelin, 59, 207n13
deities, 221n2
  as agents of reality, 222n5
  as aspects of our psyches, 222n6
  as aware, 129, 169
  and collective experience, 104–5, 106
  Hindu, 208n5
  Mesopotamian, 105, 208n5
  and nature, 105–6
  as souls, 111
  as universals and particulars, 29–30, 88
  See also God; Hinduism
Democritus, 95
Derrida, Jacques, 217–18n20, 225n5
Descartes, René, 15, 204n9
  as influence on Husserl, 213n6
  on mind-matter dualism, 40–41, 147–
    48, 150, 206n11
  reality as perceived by, 15–18, 40, 59–61,
    206n6
Deuteronomy, book of, 102, 222n7
Deutsch, Eliot, 211n14
De Wulf, M., 207n13
Dhamma (Buddhism), 215n4
dhammas (psychophysical force particles),
  76, 77, 78, 83, 85, 215n4, 216n9
dharma (Hinduism), 49
Dionysus the Areopagite, 224n25
Dirac, Paul, 154–55
disposition, concept of, 5, 53. See also intention; taṇhā (desire); values
Dōgen, 236–37n40
double-slit experiment, 160–61
Dravid, Raja Ram, 207n13, 208n4
dualist accounts of reality, 22, 23, 101, 103–4,
  115, 116, 147–48
  and free will, 109, 110

and virtue, 177
and Western monotheism, 87–88, 94
See also mind-matter distinction
Duchamp, Marcel, 162
dukkha (suffering), 71, 72, 85, 173
Duns Scotus, John, 59, 219n3, 224n25
Durgā, 54
Dvaita-Vedānta (Hindu school), 54, 211n7

Eck, Dianna, 204n8
Eckhart, Meister, 224n25
efficient cause (Aristotle), 209n18. See also
  Aristotle: on four causes
Eightfold Path, 72, 85
Einstein, Albert, 19–20, 154, 157, 160, 163,
  208n3, 230n10, 231n21
emergence, 230n2
  and awareness, 153–54
  theory of, 220n19
  weak versus strong, 152–53, 186, 230n4
emergentism, 151–53, 165, 229n14
Emerton, J. A., 223n13
emptiness, in Buddhist teachings, 83–84
enlightenment
  in Buddhism, 78–82, 168
  as experienced by the Buddha, 71
  Hindu concept of, 50–52, 79, 167–68,
    212n18
  as paradoxical, 236–37n40
Enlightenment, the (in Europe), 67, 68
entropy, 186, 237n8, 237–38n9
Eriugena, 224n25
essentialists, 205n3
evil, existence of, 110
evolution, 147, 152, 229n11
exclusivism, 8, 11, 13
existence
  and awareness, 139–40
  Buddhist concept of, 72
  Hindu concept of (sat), 47–48, 212n18
  in logic, 226n14
  and presence, 219n6
  as second-order property, 226–27n14
Exodus, book of, 107

explanation, concept of, 97, 208n3, 235n18.
　　See also Aristotle: on four causes

fanā, 176
Feibleman, James K., 207n13
Feynman, Richard, 161
final cause (Aristotle), 235n18. *See also*
　　Aristotle: on four causes
first-person accounts of reality, 21–22, 39–41,
　　43, 88, 94, 101, 103
　　Cartesian roots of, 58–61
　　as emphasizing particulars, 58, 61–65
　　Heraclitus on, 40
　　and matter, 64–65, 76, 213–14n12
　　problems with, 78, 86, 94
　　and the universal-particular distinc-
　　　　tion, 39–41
　　See also Buddhism; reality
first-person view
　　ambiguity in, 41–42
　　and awareness, 139, 183
　　as distinguished from the third-person
　　　　view, 13–15, 16–17, 19–21, 41–42,
　　　　121, 134–35, 205n4, 205–6n5
　　versus grammatical first-person, 41,
　　　　210n22, 227–28n1
　　and matter, 151
　　and the mind, 124
　　and particulars, 31, 91–92
　　and space and time, 213n10
first principles, philosophical, 39, 40–41, 57,
　　89, 101, 169, 204n12
　　in Buddhism, 75–78, 80–82, 84, 85–86,
　　　　168, 218nn25–26
　　in Hinduism, 46, 54, 55–56, 167, 210–
　　　　11n7, 212n18
　　in theism, 105, 106, 109, 112, 113, 115–16,
　　　　168
Forbes, Graeme, 207n13
formal cause (Aristotle), 206n11, 209n18. *See*
　　also Aristotle: on four causes
Foucault, Michel, 225n5
Four Noble Truths, 71–72, 75, 85, 213n7
fractal geometry, 186

free will, 7, 219n11, 223n20, 225–26n9, 234n13
　　in Buddhism, 223n14
　　in Hinduism, 223n14
　　and human agency, 108–9, 223n14
　　in monotheistic religions, 110, 223n12
　　and predestination, 109
　　and the problem of evil, 110
　　Stoic concept of, 112
　　and virtue, 177
fugue
　　as psychological disorder, 9–10, 11, 173
　　as metaphor for diverse versions
　　　　of reality, xiii, 9, 11, 166, 177,
　　　　204nn11–12, 232–33n29
Fung, Yu-lan, 208n4

Galileo, 206n8
Garden of Eden, 173, 204n14
Geertz, Clifford, 221n2, 222n5
Genesis, book of, 13–14, 27–28, 45, 107, 204n14
Gibson, J. J., 214n12
Gnosticism, 112–13, 115–16
God
　　in Abrahamic monotheism, 108–9, 116,
　　　　168, 223n13, 224n26
　　as archetype, 38, 107, 113
　　Aristotle's concept of, 112
　　and causality, 220n15, 224n29
　　as the "collective unconscious," 222n6
　　as creator, 87–88, 107, 113, 223n22
　　diverse views of, 4–5, 113
　　existence of, 38, 45, 46, 47, 102–9, 115,
　　　　220n15, 224n29, 226–27n14
　　and the existence of evil, 110
　　as first cause, 220n15
　　and free will, 109, 110, 223n12, 223n14
　　Gnostic view of, 113
　　and humans, 107–9, 113, 114–15, 116, 173,
　　　　222n9
　　ontological argument for the existence
　　　　of, 226–27n14
　　Plato's concept of, 111
　　Spinoza's perspective on, 210n4
　　union with, 224n26, 224n28

God (*continued*)
    universal aspects of, 38, 42, 87–88
    See also Brahman; religion; Western
        monotheism; Yahweh
Gödel, Kurt, 6, 156
    indeterminacy theorems of, 231n19
gods and goddesses. *See* deities
Goff, Philip, 230n4
Gold, Daniel, 221n2
Goodman, Nelson, 99, 128, 221n20
Gorgias, 40, 57, 223n20
Graham, A. C., 208n4
Greek philosophers, 206n6
    as influence on Western monotheism,
        109, 109–13
    and the universal-particular distinc-
        tion, 30–32, 112
    See also skeptics; *and names of individ-*
        ual philosophers
Greene, Brian, 208n3
Gregory of Nazianzus, 224n25
Gregory of Nyssa, 176, 224n25, 235–36n27
Grim, Patrick, 214n13
grue problem, the, 99–100, 128, 221n20
Guttenplan, Samuel, 227n19
Güzeldere, Güven, 227n19

haecceity, 219n3
Halpern, Baruch, 222n7
Hanh, Thich Nhat, 69, 79
Hartshorne, Charles, 211n10
Hebrew scriptures, 106–7
Heidegger, Martin, 207n18, 213n6, 234n15
Heisenberg, Werner, 157, 232n24, 232n26
Heraclitus, 30–31, 40, 95
Hertz, Heinrich, 157, 231n21
Higgs boson, 154, 185
Hinduism, 4–5, 27, 43–44, 70
    creation story in, 45–46, 210n6
    dual school of, 211n7
    first-person perspective in, 57, 169
    free will in, 223n14
    jñāna yoga as spiritual practice in, 52,
        55, 175

    levels of reality in, 212n18
    the mind as viewed in, 46, 124, 126
    and the mind-matter distinction,
        210n4, 210–11n7
    nondual school of, 44, 47, 50–53, 54, 56,
        57, 65, 86, 111, 167–68, 169, 210n2,
        211n8, 212n16, 212n18, 217n18, 233–
        34n2, 236n38
    philosophical difficulties inherent in,
        55–56
    philosophical schools of, 53–54, 210–
        11n7, 211nn8–9
    qualified nondual school of, 54
    and the self, 45–49, 52–54, 124, 126,
        226n13
    spiritual practices in, 49, 50–52, 54,
        167–68, 175, 212n16
    as third-person account of reality, 43–
        44, 46, 54, 56, 57
    and Western monotheism, 45, 48, 54
    See also Brahman; yogas
Hindu scriptures, 44–45
Hofstadter, Douglas R., 204n11, 236n38
Horton, Robert, 221n2
Hossenfelder, Sabine, 232n27
human beings
    as a collectivity of reference, 140,
        141–42
    as separate from God, 107–8, 113, 114–
        15, 116, 168, 173, 222n9
    See also life purpose
Hume, David, 125, 217n19
humor, 178, 179, 236n38
Husserl, Edmund
    and consciousness, 213n11, 214n12
    and phenomenology, 58, 61, 68, 212n3,
        213n6, 214n12
Huygens, Christian, 157

Ibn-al'Arabī, 113, 224–25n30
Ibn Daud, Abraham, 224n25
Ibn Gabirol, Solomon, 224n25
Ibn Rushd, 113
Ibn Sīnā (Avicenna), 224n25

idealists, 40–41
illusion
    Hindu concept of, 50–53, 55–56
    and the mind, 125
impermanence (anicca), 7
    in Buddhist teachings, 69, 71, 72–73, 85
Inanna, 105
inclusivism, 8, 11, 13, 204n12
incongruity theory (of humor), 178
individual-collective distinction, 102–3, 106,
    114, 115, 229n13
    and agency, 96, 97, 101, 103, 109, 110, 112,
        115, 121, 126, 172, 220nn16–17
    as applied to awareness, 124, 126–29,
        148, 171–74
    as applied to mind and matter, 151–53,
        158–59, 165
    and humor, 236n38
    and the metaphysics of world religions,
        166–69, 176
    philosophical puzzle inherent in,
        166–73
    and physical reality, 157–58, 160–62
    and the role of virtue, 176–77
    and set theory, 155–56
individuals and groups, 229n11
inductive logic, 99–100
instances. *See* universal-particular distinction
instrumentalism, 20, 21, 22, 29, 41, 46,
    207n16
intention, 53, 97, 123, 173, 175, 177, 214n12,
    220n15, 234n16. *See also* free will; taṇhā
    (desire); values
interdependent arising (pratītya-samutpāda),
    73, 74–75, 76, 85
    and reality as presence, 92
Ishtar, 105
Ishvara, 175, 211n13
Islam, 224n29
    and causality, 220n15
    diverse influences on, 113
    mysticism in, 176, 224n25
    See also Abrahamic religions; Western
        monotheism

Israeli, Isaac, 224n25

Jacobsen, Thorkild, 208n5
Jainism, 70
James, William, 104, 179, 234n13
Jaspers, Karl, 208n8
jñāna yoga, 52, 55, 175
John of Salisbury, 59
Johnson, George, 154–55
Johnson, Willard L., 213n7
Judaism, 106–9, 113, 115, 222n7
    God and human agency in, 223n12
    mysticism in, 224n25
    See also Abrahamic religions; Hebrew
        scriptures; Kabbalah; Western
        monotheism; Yahweh
Jung, Carl, 222n6

Kabbalah, 113, 224n25
Kant, Immanuel, 227n14, 229n15, 233n1
karma, 49, 54, 70, 177, 216n13
karma yoga, 51, 55
Kelly, Thomas, 218–19n2
Kim, Jaegwon, 219n14
Kleiber, Max, 185
koans/gongans, 174, 175, 178, 235n20
Kohák, Erazim, 212n3
Körner, Stephen, 209n17
kosmos, Greek concept of, 94–95
Krishna, 54

language
    cultural relativity of, 225n5
    universal structures in, 225n5
    See also words
language games, 204n12
Leibniz, Gottfried Wilhelm, 40–41, 219n3
Lévi-Strauss, Claude, 225n5
Libet, Benjamin, 225–26n9
life purpose, 10, 167–69, 174–79
    in Abrahamic religions, 114–15
    in Buddhism, 78–80, 217n18
    and death, 236n30
    in Hinduism, 50–53, 217n18

light
    and the double-slit experiment, 160–61
    as particle and wave, 157
    physics of, 231n21
living beings and matter, 94, 144–46, 153–54,
    228n7
Locke, John, 125, 205n5, 235n18
logic
    and awareness, 156, 225n7
    class, 36, 37
    and grue problem, 99–100
    and humor, 178
    inductive, 99–100
    limits of, 6–7, 155, 156, 166–67, 179,
        204n11, 231n15, 232n28, 233nn1–2
    many kinds of, 232n28
    modal, 231n15
    and principle of noncontradiction, 212n18
    and set theory, 155–56
    and truth, 189
    vagueness in, 232n28, 238n6
    See also paradox
logical positivism, 6–7, 203n7, 209n11,
    218–19n2
logos, 112, 115
Lord's Prayer, 102
Ludwig, Theodore M., 221n2
Luke, gospel of, 102
Luria, Isaac, 224n25
Lusthaus, Dan, 213n7

Mādhyamika Buddhism, 84, 86
Mahayana Buddhism, 83–84, 86, 217n13
Maimonides, 223n22, 224n25
manas (mind), 210n4
many worlds, theory of, 233n31
material cause (Aristotle), 206n11, 228n4. See
    also Aristotle: on four causes
mathematics, 5, 60
    and awareness, 156
    and the nature of reality, 154–56, 171,
        230n10
matter
    behavior of, 157–58

as collectivities of reference, 147–48,
    228n7
as exhibiting awareness, 161–63, 230n6
and individual agency, 220n16
lawlike behavior of, 220–21n20
and living beings, 94, 144–46, 153–54,
    228n7
material simples, 220n16
and the mind, 151, 165
publicities of, 153–54
See also mind-matter distinction; phys-
    ics; properties, material
Matthew, gospel of, 102
Maxwell, James Clerk, 157
maya (illusion), 50–51, 55–56, 177, 236n38
    five sheaths of, 52–53, 55
mereology, 209n17
Mesopotamia, deities of, 105
Mill, John Stuart, 152, 230n2
mind, the
    and the brain, 17, 134–35
    and concepts of reality, 40–42, 59–60,
        124, 171
    functional theories of, 214n13
    in Hindu philosophy, 46, 124, 126,
        210n4
    and matter, 151–52, 165
    as source of universals, 59–60
    the world as a creation of, 123–26
    See also awareness; consciousness;
        mind-matter distinction
mind-matter distinction, 13, 16–18, 40–41,
    60, 64–65, 147–48, 151, 206n6, 209n21
    Buddhist and Hindu perspective on,
        46, 52–53, 76, 210n4, 210–11n7
    See also Descartes, René; reality
moksha (liberation), 49, 114
    achieving, 50–52, 55, 79, 167
monotheism. See Abrahamic religions;
    Christianity; Islam; Judaism; Western
    monotheism
Mullā Ṣadrā, 224n25
mystery
    as aspect of reality, 162, 179

and philosophy, xii–xiii, xv, 166–69
in religion, 167
See also paradox
mystical traditions. *See* Abrahamic religions:
mystical traditions in
Nāgārjuna, 84, 204n9, 218nn24–25,
220n14, 233n2
Nagel, Thomas, 145, 235n18
naive realism (theory of perception), 206n10
Nanna (Sumerian moon god), 105
natural-artificial distinction, 94–96, 101
natural kinds, 220n19
nature, deities as agencies of, 29–30, 105–6
Neoplatonism, 113, 115–16
as influence on monotheistic religions,
223–24n25
network theory, 230n7
Neurath, Otto, 6
neurology, consciousness as understood by,
134–35
neutrino, 154
Newberg, Andrew, 235n21
Newton, Sir Isaac, 157, 207n16, 209n13
Niehoff, Debra, 229n10
nihilism, 7
nirvana, 71–72, 78–82, 83, 85, 217n18, 233n2
nominalism, 40
Nottale, Laurent, 238n11

O'Brien, Elmer, 224n25
observation
and physical laws, 20, 221n23
and scale, 184
occasionalism, theory of, 224n29
Ockham, William of, 59
Oedipus riddle, 170, 234n5
Olcott, Henry, 210n2
ontological argument, 226–27n14
Orilia, Francesco, 209n19
out-of-body experiences, 123–24, 225n1

Pāli Canon, 70, 82, 215n2, 216n9
panentheism, 211n10
panpsychism, 147, 151, 153, 165, 229n14

paradox, 164, 231n17, 231n19
and set theory, 156
Parmenides, 30, 31, 57
particles, material, 159, 160–61, 220n16, 231–
32n24, 232n26
particulars
deities as, 29–30, 88
and the first-person view of reality, 58,
61–65, 91–92
as publicities, 100, 170, 221n22
See also universal-particular
distinction
part-whole construction, 36, 37, 152, 209n17,
230n5
Patañjali, 212n20
Peirce, Charles Sanders, 219n5
Perception, 209n14
physiology of, 125
theories of, 206n10
phenomenology, 76, 88
and Buddhism, 213n7, 215n5, 218n26
Husserl's approach to, 58, 61, 68, 212n3,
213n6
philosophy
and religions, xii, 4–5, 10, 28, 166–70,
205n1
and science, xii, 6–7, 162, 203n7, 205n1
See also Greek philosophers; *and names
of individual philosophers*
photons, 157
and the double-slit experiment, 160–61
phusis, 105, 219n11, 219n13
and causality, 96
versus techne, 94–96, 101
physics, 171, 207n16
laws of, 208n3
parallel universes, 233n31
quantum field theory, 229n13,
231–32n24
quantum mechanics, 157, 159, 161, 163–
64, 165, 208n3, 232n24, 233n31
and scale, 237n7
symmetries in, 208n3
See also light; matter

Plato
    on the creator, 223n17
    on the Good, 223n16
    on souls, 111, 115, 223n15
    on universals and particulars, 31, 32, 35,
        39, 59, 112, 125, 170
play, idea of, 178
Plotinus, 113, 224n25
pluralism, 8–9, 11, 13, 41, 84
polytheism, 104–6
    and collective agency, 104–5, 115
Porphyry, 207n13
postmodern philosophies, 41, 86, 94, 190
prakṛti (matter), 54, 133, 210n4
pratītya-samutpāda. See interdependent
    arising
prayer, 174–75
predestination and free will, 109, 110
presence
    as aspect of Buddhist teachings, 80,
        82, 85
    as aspect of reality, 92–93, 102–3, 143,
        161, 177
    and awareness, 127, 129–32, 137–39, 149,
        177, 219n8
    capabilities for, 92–93, 137, 161, 165,
        219nn5–6, 226–27n14
    Derrida's perspective on, 217–18n20
    and existence, 219n6
    as ineffable, 80–82, 85–86, 135, 189
    as paradoxical, 226n10
    and publicities, 92–94, 101, 102, 130–31,
        168, 170–71, 183–84, 226n10
probabilities, as reality, 161, 165
properties, material, 64–65, 129–30, 207n14
    in Buddhism, 76–78, 80–82, 85,
        216n11
    in everyday reality, 88, 94–96, 98, 141
    in Hinduism, 52
    See also public properties
publicities
    and awareness, 135, 137–39, 153–54
    and particulars, 100, 170, 221n2
    as philosophical concept, 90–91

and presence, 92–94, 101, 102, 130–31,
    168, 226n10
    presencing of, 137–38, 149, 170–71,
        183–84
    and scale, 185
    and scientific evidence, 90–91
    truth as, 187–91
    and universals, 90–91, 98–100, 170,
        221n22
public properties, 63–65, 88
    and awarenesses, 128, 129–32, 135, 171
    as mind-independent, 103–4
    as shared experience, 103–4, 128–29, 140
    as universals, 89–91
    See also properties, material
Pure Land Buddhism, 83, 84, 114–15
puruṣa, 54, 133, 211n15
Pyrrho, 40, 57, 204n9

qualia, and the first-person versus third-
    person views of reality, 81
quanta of energy, 157, 158
quantum computing, 163–64
quantum field theory. See physics: quantum
    field theory
quantum mechanics. See physics: quantum
    mechanics
Qur'an, 102, 113
    God and human agency in, 223n12

Rāja Yoga, 51, 212n20
Raju, P. T., 211n7
Rāma, 54
"real"
    and collectivities of reference, 238n5
    as distinguished from "true," 187–89
    meanings of, 18–19, 29, 40, 203n1,
        206–7n13
    See also reality
realism, 20, 21, 22, 29, 40–41, 209n21
reality
    alternate paths for, 233n31
    and aware beings, 94–96, 169–70, 211n7
    as capability for presence, 92–93

Cartesian gap in our understanding of,
19–21, 42, 67, 106, 128–29, 134–35,
147–48, 150
causality as aspect of, 65, 93
and collectivities of reference, 140–49,
152–53
composite, 163–64
concept of, 18–21, 205n2, 206–7n13
and counterfactuals, 93
definition for, 89, 94, 137, 203n1
deities as agents of, 104–6, 222n5
Descartes's perspective on, 15–18, 59–61
diverse views of, 4–5, 7–9, 11, 12–13
dualist approaches to, 22, 23, 87–88, 94,
101, 103–4, 110, 115, 116
first-person versus third-person versus
dualist accounts of, 13–15, 16–17,
19–21, 22–23, 41–42, 88, 94, 101,
103, 106, 121, 151, 166, 166, 207n19
as fugue, 9, 11, 166, 177, 204nn11–12,
232–33n29
and humor, 178, 179
as information, 232n26
mathematics as applied to, 154–56, 171,
230n10
and the mind, 40–42, 59–60, 124, 203n1
as mind-independent, 18, 21, 29, 59, 61,
87, 205n2, 206n10
in monotheistic religions, 87–88, 107
mystery as aspect of, 167, 179
and parallel universes, 233n31
and presence, 80–82, 91–94, 102–3, 143,
161, 177
as public, 89–91, 103–4, 143
scale as dimension of, 183–86
and science, 5, 20–21, 154–64
and spiritual practices, 174–76
as the third-person view, 18–19, 21
in time, 171, 213n10, 233n30
as "true for all," 3, 68, 203n1
and truth, 187–89
Western concept of, 18, 40
See also Abrahamic religions; awareness; Buddhism; first-person

view; Hinduism; publicities;
third-person view; universal-
particular distinction
reason, Stoic concept of, 112
Reese, William L., 211n10
religion
and community, 176, 179
disbelief in, 205n16
mystery as aspect of, 167, 173–74
and the nature of reality, 8–9, 166–70
and philosophy, 4–5, 10, 28, 166–70,
205n1
as philosophy in practice, xii, 4
as psychological disorder, 4
scripture as sole source of authority, 4,
205n1
See also Abrahamic religions; Buddhism; Hinduism; spiritual practices; Western monotheism
religious diversity, 169
approaches to, 8–9
representative realism (theory of perception),
206n10
Rey, Georges, 228n6
Ricoeur, Paul, 213n6
Robinson, Richard, 213n7
robots, as distinguished from human beings,
144–45, 228–29n9
Rorty, Richard, 206n7, 209n19, 225n5
Rosenberg, Alex, 207n16, 219n7
Rosenthal, David, 133
Russell, Bertrand, 227n14
paradox proposed by, 156

Sāṃkhya (Hindu school), 54, 133
samsara
in Buddhism, 72, 173, 235n19
in Hinduism, 48, 49, 55
Sangha, 215n4
Śaṅkara, 236n38
Sarvāstivāda Buddhism, 217n13
scale
as dimension of reality, 183–86, 237n3
and Newtonian physics, 237n7

scale (*continued*)
  and thermodynamics, 186
  and time, 184–85, 237n6
schizophrenia, perception of reality as
    affected by, 3–4, 123
Schoedinger, Andrew B., 208n4
Schopenhauer, Arthur, 217n15
Schrödinger, Erwin, 158, 159, 163
Schrödinger's cat, 163–64, 165
science
  limitations of, 5–7, 10, 11, 171
  and philosophy and religion, xii, 6, 7,
      162, 203n7, 205n1, 212n18, 233n2
  realism versus instrumentalism in the
      philosophy of, 20
  reality as conceived by, 5, 20–21, 154–64
  See also cells; mathematics; physics
Searle, John, 228–29n9
self, the, 212n3
  Abrahamic concepts of, 109–13, 124,
      126, 226n13
  alternate accounts of, 226n13
  awareness of, 67
  Buddhist concepts of, 72–73, 124
  Hindu concepts of, 45–49, 52–53, 124
  Greek concept of, 115–16
  and the mind, 226n13
  and personal identity, 148–49, 150
  in space and time, 148–49
  See also anatta (no-self); Brahman;
      consciousness; human beings
sense experience and universals, 6, 16, 31, 35,
    42, 59
set theory, 155–56, 231n17
  and awareness, 156, 171
  See also class logic
Shaner, David Edward, 213n7
Shema, 102
Shinto, 5
Shiva, 44, 54
Siddhartha Gautama. *See* Buddha, the
signifiers and the signified, 146–47
skeptics, 40, 57, 112, 204n9, 206n6
Skinner, B. F., 203n3

Skinner box experiment, 4, 203n3
Smith, Huston, 212n20
Sophists, 40
sortals. *See* natural kinds
soul
  as distinct entity, 222n9
  in Gnosticism, 113
  as life spirit, 107
  Plato's concept of, 111, 112, 115, 223n15
  See also mind, the; self, the
space
  first-person versus third-person, 213n10
  measurement of, 164, 184–85, 233n30
  See also scale
Sperling, S. David, 222n7
Spinoza, Baruch, 210n4
spiritual practices, 174–77
  and the development of virtue, 176–77
  and our understanding of reality,
      174–76
  See also Abrahamic religions; Bud-
      dhism; Hinduism
Stein, Gertrude, 208n1
Stoics, 112, 115
Strawson, Galen, 229n14
string theory, 186
subjectivity, 134, 145
substances
  Aristotle's concept of, 31–32, 111–12, 115,
      215n7
  primary as distinguished from second-
      ary, 32
suffering
  as aspect of Buddhist teachings, 75, 85
  and everyday life, 173
  and the Four Noble Truths, 71–72
  and humor, 179
śūnyatā (emptiness), 83, 84, 85, 218n24
Śūnyavādins, 83
Swoyer, Chris, 209n19

Tagore, Rabindranath, 43
Tanakh, 106, 107, 223n13, 223n22
taṇhā (desire), 71–72, 79, 85

Tarski, Alfred, 189
tathatā (suchness), 84, 85–86, 92
Taylor, G. I., 160
techne versus phusis, 94–96, 101
teleological, universe as, 235n18
Teresa, Mother, 52
Thales, 30, 39, 95
theism. *See* polytheism; Western monotheism
Theravada Buddhism, 83, 84
thermodynamics, 186
things in themselves, 205n5
third-person accounts of reality, 21–22, 39–41, 43, 88, 89, 94, 101, 103
  critics of, 57
  problems with, 55–56, 94
  and the universal-particular distinction, 39–41
  and universal truths, 41
  See also Hinduism; reality
third-person view
  as distinguished from the first-person view, 13–15, 16–17, 19–21, 41–42, 121, 134–35, 205n4, 205–6n5
  and first-person data, 134–35, 136
  and presence, 91–94
  and publicity, 89–91
  as reality, 18–19, 21
  and space and time, 213n10
  and the universal-particular distinction, 27–29
Thomas Aquinas, 59, 211n9, 220n15, 224n25
Tibetan Buddhism, 218n24
time
  first-person versus third-person, 213n10
  measurement of, 184–85, 233n30
  and personal identity, 148–49
  and scale, 183–84
Torah, 106–7
transmigration (cross-species reincarnation), 4, 48
tripuṭi, 211n8
"true," as distinguished from "real," 187–89.
  See also truth

truth
  in Buddhism, 233n2, 233n2
  context as factor in, 238n6
  correspondence theory of, 187–89
  definition for, 187–88
  as distinguished from "real," 187–89
  in Hinduism, 212n18, 233n2
  overview of theories of, 189–91
  as publicities, 187–91
  as relative, 205n2
truths, universal. *See* universal truths
types and tokens, 28, 29, 208n1

Ulin, Robert Charles, 221n2
unconscious awareness, 66, 67, 124, 127, 132, 214nn–17, 228n2
universal-particular distinction, 27–29, 42, 221n22, 234n7
  ambiguous relationship of, 38–39, 42, 170–72
  Aristotle on, 31–32, 39, 111–12
  atoms as example of, 32–35, 38
  compositions of, 36–37
  historical roots of, 29–32
  and the nature of reality, 39–42, 87–88, 172
  as principles of nature, 32–38, 47
  See also individual-collective distinction; particulars; universals
universals, 209n15, 223n22
  in Buddhism, 75, 86
  and collective agency, 97–98
  deities as, 29–30, 88
  in Greek philosophy, 94–95
  in Hinduism, 47, 57, 85, 211n9
  in Indian philosophy, 208n4
  presence as, 82
  as publicities, 98–100, 170, 221n22
  public properties as, 89–91
  See also reality; universal-particular distinction
universals, mind-independent, 29, 125, 207n13, 219n13
  challenges to, 59–60, 68, 125

universal truths, 5, 21, 39–40, 41, 205n1
  and the first-person versus third-
    person view of reality, 169
universe
  mathematics as applied to the under-
    standing of, 154–56
  as having essential properties, 220n19
  scale as dimension of, 183–86
Upanishads, 44–46, 47, 55, 70, 210n6

vagueness. *See* logic: vagueness in
values
  as an aspect of awareness, 173
  as an aspect of maya (illusion), 53
  as distinguishing yogas, 51, 55
  and facts, 17
  as properties, 235n18
Vasubandhu, 218n26
Vātsīputrīya Buddhism, 217n13
Vedas, 44–45
vijñāna (conscious awareness), 47, 67, 76. *See
  also* awareness; consciousness
vipassanā meditation, 175
virtue
  and the individual-collective connec-
    tion, 176–77
  and Plato, 111
  as viewed in Hinduism, 49
Vishnu, 44, 51, 54, 210n6
Vision, Gerald, 230n4
Viśiṣṭādvaita-Vedānta, 54
Vivekananda, Swami, 210n2, 212n20

wave mechanics, 159, 160–61, 231–32n24,
  232n26

Wehner, Stephanie, 232n26
Weinberg, Steven, 10
Weinfeld, Moshe, 222n7
West, Geoffrey, 185
Western monotheism
  and dualist accounts of reality, 87–88,
    102–4
  early texts of, 106–9
  and free will, 108–9, 110, 223n12
  Greek philosophers as influence on,
    109, 111–13
  and Hinduism, 45, 48, 54
  and humans' separation from God,
    114–15, 168
  See also Abrahamic religions; God
Whitehead, Alfred North, 219n5, 234n16
Wigner, Eugene, 155
William of Champeaux, 207n13
Wittgenstein, Ludwig, 204n12, 229n12,
  234n14
Woodhouse, Roger, 205n5
words, as artifacts, 146–47, 150, 229n12. *See
  also* language
Wright, Robert, 221n2, 222n7
Wu-men Hui-k'ai, 235n20

Yahweh, 106–9, 222n7, 222n9. *See also* God
Yoga (Hindu school), 54, 133
Yogācāra Buddhism, 84, 86, 213n7, 218n26
Yogananda, Paramahansa, 210n2
yogas, in Hindu spiritual practice, 51–53, 176
Young, Thomas, 157

Zen/Chán Buddhism, 83, 218n24
Zen koans. *See* koans/gongans